Learning AI Tools in Tableau
Level Up Your Data Analytics and Visualization
Capabilities with Tableau Pulse and
Tableau Agent

Ann Jackson

O'REILLY®

Learning AI Tools in Tableau

by Ann Jackson

Published by O'Reilly Media, Inc., 1005 Gravenstein Highway North, Sebastopol, CA 95472.

O'Reilly books may be purchased for educational, business, or sales promotional use. Online editions are also available for most titles (*http://oreilly.com*). For more information, contact our corporate/institutional sales department: 800-998-9938 or *corporate@oreilly.com*.

Acquisitions Editor: Michelle Smith	**Indexer:** BIM Creatives, LLC
Development Editor: Corbin Collins	**Interior Designer:** David Futato
Production Editor: Ashley Stussy	**Cover Designer:** Karen Montgomery
Copyeditor: nSight, Inc.	**Illustrator:** Kate Dullea
Proofreader: Sharon Wilkey	

January 2025:   First Edition

Revision History for the First Edition

2025-01-13:   First Release

See *http://oreilly.com/catalog/errata.csp?isbn=9781098175788* for release details.

978-1-098-17578-8

[LSI]

Table of Contents

Preface

When I decided to write this book, I wanted to give you a chance to understand the history of Tableau. I also wanted to give you a grounding and understanding of what artificial intelligence (AI) is. And finally, I wanted to address the impacts of this new technology, not only in analytics but also in the world. My goal in these decisions is to ensure that we are on the same page throughout the book, we are speaking the same language, and we're addressing some important issues early on so that they aren't neglected. Often I've found in our profession, the lack of saying things leads to confusion or the prolonging of important discussions, so instead, let's get started.

A Brief History of Tableau

Tableau was founded in 2003 as a by-product of a Stanford research project and dissertation. The key innovation behind the software is VizQL, a visual query language that allows those working with data to drag and drop fields in data sets on a canvas to query and answer questions. It's a very intuitive system because all in-depth coding and manipulation of the data is done within the VizQL engine, and the person dragging and dropping gets direct visual feedback in the form of charts and graphs. I was first introduced to this technology around 2015, and to date I haven't seen a more intuitive way to analyze and work with data.

In 2007, the software expanded to include Tableau Server—a way to host and publish visualizations to true audience members and end users (not just those who create the charts)—and Tableau Reader, a free application that allowed anyone to access and open Tableau files, called *workbooks*, locally. This improvement paved the way for what is a big part of modern analytics, the idea of *self-service analytics*. With the inclusion of these two products, now someone could build charts and share them with anyone who was looking to use data as an aid in decision making or understanding. The responsibility and cycle of reaching out to a data analyst for answers to data-related questions was shifted back to the end user, with the goal that they become more empowered and able to answer their own questions, to explore the data, and to know it more intimately.

From 2007 to 2016, Tableau continued to improve upon its core product, Tableau Desktop. Some of the most powerful innovations at the time were the inclusion of connectors to a myriad of databases and data sources. Data professionals using the tool could connect to Microsoft Excel spreadsheets with hand-constructed budget numbers and could also connect to larger databases holding transactional data. To this day, Tableau continues to be unique in the data tools landscape because it takes what I like to call a *data agnostic* approach. Most business intelligence (BI) tools thrive on connecting to specific data sources (like Microsoft's legacy SSRS and SSAS) within an application stack (Microsoft), somehow making the visual analytics component the last step in the process. Tableau upends this paradigm by taking the data as it is, wherever it is, squarely aiming to make the job of the data analyst easier: easier to access data, analyze it, understand it, and disseminate it.

Beyond the continued improvements in data connectivity, Tableau added in the ability to construct *level of detail (LOD) expressions* in version 9. This gave Tableau visualization creators the ability to define unique aggregations for calculations independent of the aggregations and fields represented in a chart. I started using Tableau right before this feature was released, so I have a very strong sense of what was possible and the impact of this improvement. Suddenly, deeply complex queries and comparisons could be made within data sets, all in the same chart. At the time, I was working as an analyst in healthcare, and I remember being amazed at how I could now show zoomed-in data, like the best-performing provider, against a state, regional, and national average—all in one chart.

It was also around this time (2016) that Tableau released version 10. This introduced a modern user interface, a bespoke Tableau font, and the ability to relate, both by unions and more importantly by joins, tables across different databases. Suddenly, as an analyst, you could more easily enrich data that was disparate. Combining information from multiple systems without the prerequisite of a data warehouse or single-source repository was now possible (for die-hard Tableau users, data blending has been available since 2010, but with extreme limitations). Requests I made as an analyst changed from asking IT for constructed data sets to asking for direct access to databases. The unbelievable power of accessing the raw data was exciting. As an analyst, this access gave me a better understanding of the applications used for the business: I could now understand how it was really organized and what data was really there. It filled in a lot of gaps of whether certain types of analysis were possible.

From this period of time to the end of 2019, Tableau started addressing two new problems: data access, storage, and management; and analytics tools beyond visualizations and dashboards. This shift was a significant milestone, quietly signifying that its visual analytics product was now quite mature and there was now space to spread into other areas of enterprise analytics.

On the data access, storage, and management front, Tableau introduced Prep Builder (its data preparation tool), a new proprietary data storage format (Hyper), and eventually a data catalog. All these innovations represented massive improvements to the quality of life of an analytics team working with data. Teams were now getting legitimate data preparation tools that rivaled SQL in terms of query and cleaning capabilities with the intuitive and visual interface Tableau is known for. Hyper, the name for Tableau data extracts, allowed underlying data sets to be larger, with both more rows and columns. All analytical operations on dashboards became faster. Data started to get less aggregated underneath the dashboards, allowing for analysis at both the macro (executive) and the micro (data analyst) levels, all in the same space and at the same time. It was now easier to point to the exact record of data that was contributing to a monthly trend line. The data catalog addition helped legitimize Tableau Server as an enterprise-level BI platform—marking the change of Tableau from being categorized as a pretty dashboarding tool to a more complete BI solution.

As I mentioned, it was also during this time that Tableau started working on two innovations unrelated to visual analytics:

Ask Data
A natural language querying tool (released in 2019)

Explain Data
A statistical analysis tool that surfaced interesting statistical facts as the user created charts

Looking back on this now, it is clear to see that these innovations were a precursor to AI, but at the time it was a little bit harder to see it that way. The introduction of Ask Data introduced a lot of anxiety to data practitioners. Previously, they had been able to control and check that any questions an end user asks of a dashboard would surface factually accurate answers. But with a language interpretation engine spitting out figures to someone less familiar with how to analyze data, it was tough to trust.

With Explain Data, the statistical figures that surfaced may or may not be interesting, and for the creator of data products, that information felt repetitive. For dashboard consumers, it felt like Explain Data would, similarly to Ask Data, introduce facts (albeit statistically and factually accurate) that might or might not be of significance. Both do push the bar of self-service analytics more toward the end user. However, they also invited an erosion of trust (with their audiences) that data professionals had been building up through highly developed and iterated dashboards.

From the end of 2019 onward was what I call the *Salesforce era*—maybe as a nod to Taylor Swift, but more as an explanation of everything that has been the focus of Tableau since its acquisition by its parent company, Salesforce, in the second half of

2019. Since then, Salesforce and Tableau have turned their focus almost exclusively to the inclusion and integration of Salesforce and Tableau technologies and products. On the data analysis side, it meant the inclusion of Salesforce's Einstein, a mostly predictive and scoring analytics technology, allowing for more advanced analysis in both Tableau Desktop and Prep Builder. Improvements were also made in connecting to Salesforce data sources and embedding Tableau visualizations in its customer relationship management (CRM) application. And finally, analytics expanded beyond CRM and Tableau Server/Cloud and into Slack, the communications platform Salesforce acquired in December of 2020. The Einstein integration and outbound communication to Slack were released in 2021. Data Stories (summarized text insights based on charts) released in 2022. Figure P-1 shows a timeline of these innovations.

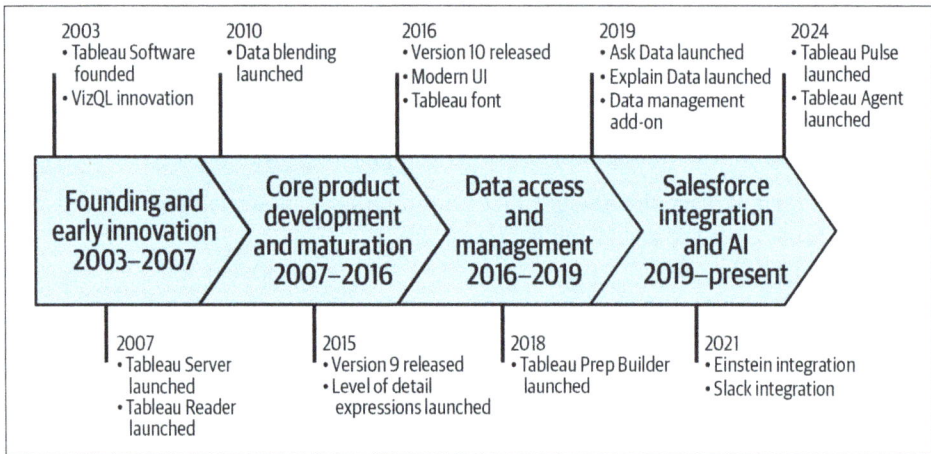

Figure P-1. Timeline of major Tableau innovations

Looking back on this history lesson, it's easy to see that Tableau has always positioned itself at the forefront of innovation in its mission of "helping people to see and understand data." Tableau has not only expanded what it means to have access to data and user-friendly tools, but also introduced more features to lower the bar of entry into a data-informed workplace.

Introducing Tableau Pulse and Tableau Agent

And now we have reached the current time and the release of technologies that are at the heart of this book: Tableau Pulse and Tableau Agent (formerly Einstein Copilot).

Tableau Pulse

Tableau Pulse is a completely new interface to and component in the Tableau product line. As of this writing, Pulse exists only within Tableau Cloud, Tableau's software as a service (SaaS) offering. Within this interface, important business measurements called *key performance indicators* (KPIs) can be constructed and are served up in a separate area for end users to *follow*, much like following someone on a social media platform. Pulse includes a Metrics layer where data professionals curate different measurements and give definition to them. Metrics are served up as simple and uniform charts along with AI-generated summaries, called insight summaries. End users can subscribe to receive email or Slack communications including these summaries at a given frequency. They can also access their followed metrics within the Tableau Mobile app, a pared-down version of Tableau Server/Cloud available on mobile devices, or in a browser. In this web experience, they can access simple filters and ask questions in a conversant manner, expanding their ability to interact with the KPIs.

Chapter 2 introduces Pulse in more detail, including how to get started and work with it. You'll also see hands-on examples based on industry use cases in Chapter 5.

Tableau Agent

Tableau Agent is an AI assistant located directly within the Tableau Cloud products. Its purpose is to help the data professional analyze data and build charts more quickly. Tableau Agent accomplishes this speed enhancement by jumpstarting the user with key questions they may want to ask of their data. It also provides chart creation and calculation support. In theory (and practice), Tableau Agent can help get you up to speed more quickly with a data set and create charts through conversation. All of this aims to take the iterative and feedback-driven nature of the Tableau platform and make it even faster for the data worker to find, process, and share impactful information.

Chapter 7 discusses Tableau Agent, demonstrating its capabilities and providing you with prompt best practices. Finally, Chapter 8 discusses some ways I imagine this technology expanding and supporting your work.

Chapter 1 begins to get into the weeds of the how, why, and when to use these technologies by first helping you understand the technologies that power these tools. That's also where you'll begin to learn about the risks and considerations you must acknowledge before you start implementing and massively sharing these innovations.

Conventions Used in This Book

The following typographical conventions are used in this book:

Italic
    Indicates new terms, URLs, email addresses, filenames, and file extensions.

Constant width

> Used for program listings, as well as within paragraphs to refer to program elements such as variable or function names, databases, data types, environment variables, statements, and keywords.

Constant width bold

> Shows commands or other text that should be typed literally by the user.

 This element signifies a tip or suggestion.

 This element signifies a general note.

 This element indicates a warning or caution.

O'Reilly Online Learning

 For more than 40 years, *O'Reilly Media* has provided technology and business training, knowledge, and insight to help companies succeed.

Our unique network of experts and innovators share their knowledge and expertise through books, articles, and our online learning platform. O'Reilly's online learning platform gives you on-demand access to live training courses, in-depth learning paths, interactive coding environments, and a vast collection of text and video from O'Reilly and 200+ other publishers. For more information, visit *https://oreilly.com*.

How to Contact Us

Please address comments and questions concerning this book to the publisher:

O'Reilly Media, Inc.
1005 Gravenstein Highway North
Sebastopol, CA 95472
800-889-8969 (in the United States or Canada)
707-827-7019 (international or local)
707-829-0104 (fax)
support@oreilly.com
https://oreilly.com/about/contact.html

We have a web page for this book, where we list errata, examples, and any additional information. You can access this page at *https://oreil.ly/learning-ai-tools-in-tableau-1e*.

For news and information about our books and courses, visit *https://oreilly.com*.

Find us on LinkedIn: *https://linkedin.com/company/oreilly-media*

Watch us on YouTube: *https://youtube.com/oreillymedia*

Acknowledgments

I want to thank my husband and life partner, Josh Jackson. A lot has changed over the years, but one thing has stayed the same: your unconditional love and support. Who knew that two kids in a physics chat room could create such a beautiful life together?

To my close family and friends, thank you for your patience and understanding throughout the writing process of this book. I'm excited to emerge from my writing cave and resume more board game nights, as well as replying to your texts in a time-lier fashion. Wordle Geniuses, your daily commitment to our group chat has kept me sane the last two months of writing.

To my technical editors, Lorna Brown, Jami Delagrange, and Josh Jackson (yep, same guy!)—thank you for your feedback, comments, and encouragement. I'm grateful to have three people who are willing to take time out of their lives to ensure this project meets my ridiculously high expectations.

To my data kittens, thank you for inspiring me and going on travel adventures with me. I've enjoyed every moment outside of the classroom we have spent together and look forward to visiting Denver (and yes, seeing you all again in NYC).

Finally, to the AI researchers and innovators who have progressed the technology featured throughout this book, thank you. You've breathed new life into my passion for technology and analytics. I look forward to the many transformations yet to come.

CHAPTER 1

AI in the Tableau Platform

Artificial intelligence isn't new. It's been around since the famous mathematician Alan Turing first asked, "Can machines think?" in his well known 1950 paper "Computing Machinery and Intelligence." AI was formalized into an academic area for research, study, and innovation (according to many academics and historians in the field) in 1956. In the beginning, we can imagine that the goals of AI were to replicate or emulate human intelligence and decision making. This may have been best stated by computer scientist John McCarthy, who coined the term *artificial intelligence*: "Every aspect of learning or any other feature of intelligence can in principle be so precisely described that a machine can be made to simulate it." Many would add that computers could (in theory and now in practice) handle complex operations—sometimes much more complex (particularly mathematical ones) than humans could.

Since that time, AI has slowly woven itself into every area of our lives. The idea of being bested by a computer at chess is nothing new. In fact, AI's place in any sort of strategic game now feels common and expected. More recent and prolific applications of AI would include recommendation engines, such as those determining what to watch next on Netflix or what to buy on Amazon. Voice assistants like Apple's Siri that can translate your speech into some type of action also qualify as AI. And then there are self-driving vehicles like Google's Waymo that take the idea of *autopilot*, a concept that's been around for roughly 100 years, to a brand-new paradigm.

So much of AI has depended on the capabilities of computers. Early computers were limited in the information and knowledge that they could hold. And similarly, even if they had access to the knowledge, the computational and processing power necessary to access, retrieve, and serve up that information was massive. But in the early 2000s, the pace of innovation with computer hardware caught up with the needs of AI. Multiple processors and multithreading proliferated, smartphones became prominent fixtures in our everyday lives, and cloud computing became the new normal.

AI in Analytics

As advancements in computer hardware became more prominent, the domain of analytics started to see its own advancements. In particular, this meant the inclusion of more advanced methods of mathematical analysis. *Multivariate analysis* became more pervasive in business applications as more knowledge workers could lean on computers to help with the complex mathematical computations. *Predictive analytics* and the ways in which something could be predicted expanded from the simple linear regression model of old to much more robust and sophisticated methods. Classification methods, like *k-means clustering* (available in Tableau) were possible. Machine learning (ML), both *supervised*, where the input and output are provided to the model, and *unsupervised*, where models are given the freedom to derive their own patterns, came into prominence. A computer could now be given a massive amount of data and draw conclusions with varying levels of aid and decision input from humans.

Then came *natural language processing* (NLP). As mentioned in the Preface, Ask Data was released by Tableau in 2018, allowing users to ask a question and get an answer, but speech recognition and transcription tools have been living in the business world for much longer. *Sentiment analysis*—or the process of scoring language to determine whether it is positive, negative, or neutral—started becoming mainstream. Computer languages like Python and R became prominent, and the practice of data science and the emergence of the data scientist reached a fever pitch in the 2010s, along with the term *big data*.

2018 also heralded the first *generative pre-trained transformer* (GPT) technologies, a type of *large language model* (LLM), by OpenAI, the company that soon became known for its famous ChatGPT chatbot, capable of having coherent and improvised conversations with humans, and its generative digital art tool DALL-E. I was introduced to both technologies in 2022. I signed up for the early invitation-only beta of DALL-E and loved creating digital art that would have been previously out of my own reach (like melting bar charts in the style of Salvador Dali, and Darth Vader walking Princess Leia down the aisle as a heartfelt watercolor vignette (see Figure 1-1). AI-generated sonnets, poems, parodies, and rote business communications started popping up as novelties at my workplace. My then-boss jokingly posted his ChatGPT-created professional biography which included glowing praise of who he was but was also littered with inaccuracies and accolades that he had never received. Generative and creative AI had officially arrived.

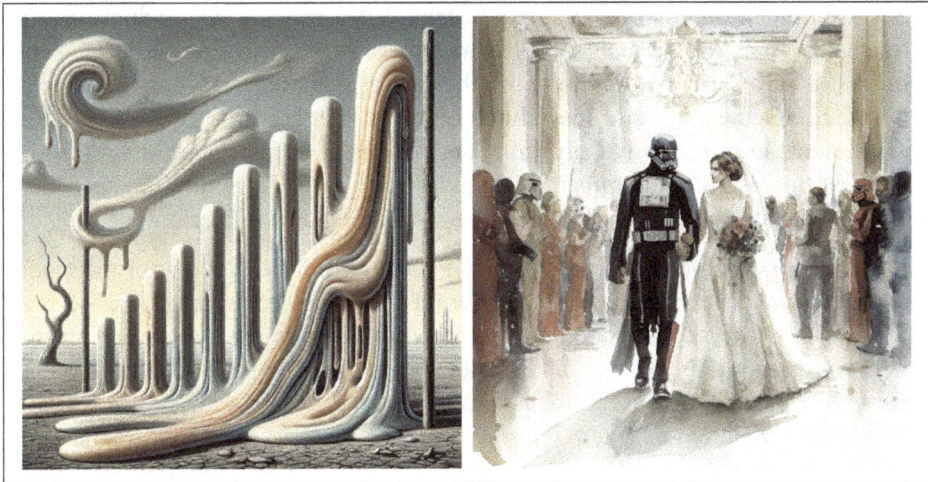

Figure 1-1. AI-generated art created using DALL-E (left) and MidJourney (right)

The technology and structure underpinning LLMs and GPTs are in full use in both Tableau Pulse and Tableau Agent, the focal points of this book. Both rely heavily on LLMs. In Pulse, an LLM is used to summarize the insights and serve them up in a meaningful way. With Tableau Agent, you get direct interaction and results based on your input into a chat box. This is a very privileged position for technology to have in our domain, and as such, it's important for analytics professionals to understand how AI technology works as much as possible. Referring to generative AI as opaque or enigmatic technology isn't acceptable when you're serving up information to end users to make business-critical and sometimes life-or-death decisions based on data.

Generative AI Explained

A good way to think about how LLMs and GPTs function is to start by thinking about words as a set of coordinates, called *word vectors*. Similar to latitude and longitude, imagine that each word has an elaborate set of coordinates determining where it is located in the universe of words. Unlike the confines of the three dimensions that we experience in the physical world, each word in the universe of words has a very large number of dimensions (300+). Moreover, each word can be represented multiple times in the universe based on what it represents. Good examples of this are *homonyms*, words that are spelled the same but have different meanings, like the word *hide*, which can mean an animal skin pelt and can also mean to position something out of sight. It's easy to imagine that there are two separate coordinates for these two concepts, just as there are (at least) two definitions of *hide* in a dictionary.

LLMs are trained on copious amounts of text, called *training data,* which they go through to systematically assign coordinates to each word in the universe. Those coordinates have a natural proximity to other words that are closely related, and you can further imagine that those coordinates can shift slightly with each additional new passage of text received. Eventually, from a mathematical angle, there are diminishing returns on this—similar to limits in calculus: as you approach infinity, the coordinate of a word eventually settles at a final location within the word universe.

Alongside the coordinates of words is the GPT, or transformer, process of the model. I like to explain this process in terms of *layers.* These layers (for example, GPT-3 from OpenAI has 96 layers) go through the text input called a *prompt* it receives, categorizes the words within the prompt, and eventually aims at comprehending it. All of this is somewhat opaque, because humans haven't dictated to the models at which layer different actions should be taken. From what has been observed, it appears that most models start by understanding the sentence syntax and role (noun/verb/adjective/pronoun) each word takes on. After those initial layers are done, more contextual layers process the prompt. I try to think of this as what humans innately and effortlessly do. When you start reading a book, you get a grounding in the words, and eventually, once you've read enough, your brain starts imagining the whole scene. Figure 1-2 demonstrates how an LLM may process a simple sentence through the transformer layers.

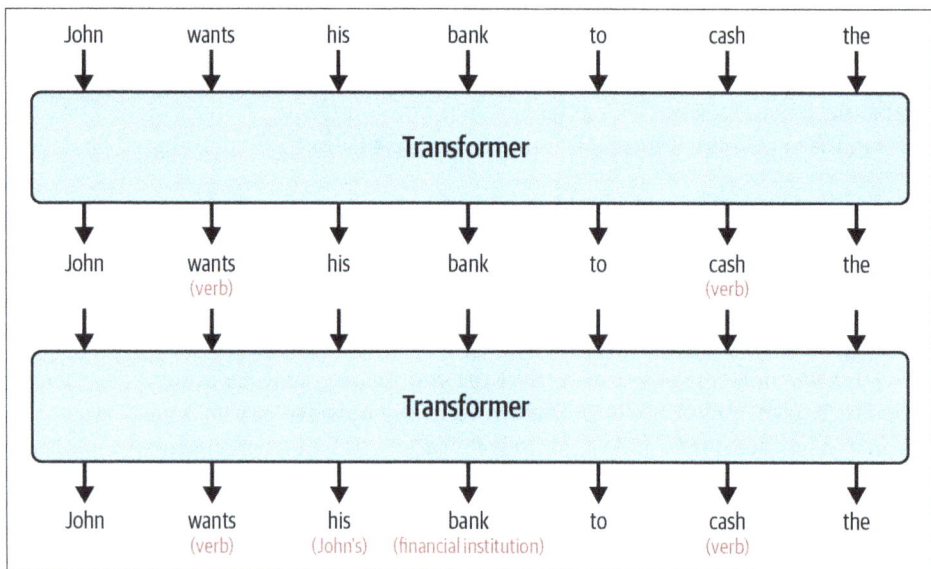

Figure 1-2. An example of how an LLM processes a sentence

After the prompt is processed comes the heart of the model's generation: a prediction of what should come next. Again, this is not unlike what humans might also innately do on their own. If someone asks you what your favorite ice-cream flavor is, at the very least you are limited by the list of flavors you are aware of (possible predictions). In addition to that list, you rely on your own experiences with each of these flavors, and that eventually leads you to a single answer. Maybe it's hard to land on one flavor, or you might be mentally scoring each flavor to get to a result. Maybe vanilla is your favorite "safe" flavor because it's hard to get vanilla wrong, but you experience more pleasure and delight when eating mint chocolate chip. Maybe Ben and Jerry's Cherry Garcia is your guilty pleasure, but not every ice-cream parlor has that on the menu. It is possible to answer the question differently depending entirely on the question's *context*: who is asking, your recent ice-cream experiences, where you are, what you are craving. The models resolve this by relying on interpreting the context of the prompt and what they have learned from the training data. And it's important to say, the model's training data is far beyond the scope of what any one human knows, so it quite often has a very broad understanding of the possible answers.

Remember, the whole act of processing the syntax and sentence structure of the prompt through comprehension and context is completely dependent on the information the model has on hand. The predictions it is able to generate are from the information it's been given, which has been presented in sentences written by humans. So in particular, when we apply LLMs to a narrow field like coding or analytics, the model is relying entirely on what humans have already done. This isn't to say that there aren't original outputs (dare I say thoughts) or creativity in the process. In fact, one of the reasons I like generative AI for creating art is that it is not bounded by practicality. DALL-E may generate beautiful images that include an extra hand or misspelled word, because the prediction model has resolved to output the "best result." A human artist would never paint an extra limb unless it was highly intentional, even if it better conveyed the idea behind the overall piece of art.

Given how LLMs function, we must treat their prediction method as a double-edged sword. It can be contaminated by human bias innate in the text, limited by the information it has been given, and may predict an output that is nonsensical or factually inaccurate. It may also provide answers or results that are uniquely different from what a human (or meticulously crafted algorithm) completing the same task may produce.

Risks and Considerations with AI

There are three big areas of risk and consideration I want to discuss:

- Model bias
- Model hallucination
- Worker displacement

It's important to discuss these because they will inevitably come up on your organization's journey to include generative AI tools in your analytics practice. Knowing what these issues are up front arms you with the knowledge your organization needs to make responsible and informed decisions about these tools.

Model Bias

You've already learned that a model can be biased based on the information it has been trained on or has access to. But a model can also be biased based on how the algorithm was designed. A relatively easy example to understand is an algorithm that has been tuned to over-rely on information that shows up frequently or more recently. When that is applied to the question *Who is the most popular singer?*, the result might be *Taylor Swift* based on the success of her recent *Eras* tour and how often her music is requested and served up. But looking at the question over a longer span of time could surface a different answer: *Frank Sinatra*, who released an astounding 59 studio albums over his 50+ year career. You can imagine how that question gets murkier when you're applying it directly to data or statistical information. The question *What has caused the decline in sales?* (after looking at a chart of trending sales values) could yield different answers based on whether the model considers only the data points observed in the chart or has access to the entire history of sales.

User interaction is another tricky part which will naturally influence what the model produces. If the model knows that you are the manager of the electronics department or if you've asked many questions that tilt toward electronics, it will likely produce an output more directly related to electronics. This could produce an answer that is accurate, but perhaps not the largest or most direct root cause of the example question *What has caused the decline in sales?* I often see this unfold in my own interactions with ChatGPT, where I'll ask for something in a list format, and from then on it has trouble breaking out of a pattern of listing items, even if my follow-up requests aren't best resolved with lists. ChatGPT also tends to use similar language or structure to what I am providing, which, again, seems logical (we humans often tend to adopt the same words and style of those around us) but works against getting an unbiased response.

Finally, there are social prejudices, stereotypes, and representation biases to contend with. These can appear by a model parroting back the most popular prejudices that have woven their way into the body of human knowledge. A fairly benign example is the word *nurse*. When you read that word, your mental image of a nurse is likely a woman. But you could extrapolate that act out to something more analytically oriented. For example, you seek help in creating a formula to predict how sales will perform, and the model defaults to recommending a linear regression because of its popularity in analytics. In a nonharmful way, the model may simply be providing the most common way to address your data. However, the model's response could also be negating a more sound and reasoned method that would be more appropriate for addressing your data, leaving you with a method that may not be the right fit for your situation.

Model Hallucination

Model hallucination is a model's return of a result that is not factually accurate or not grounded in reality. The example of the extra limb added to a generated image of a human figure mentioned earlier comes to mind, or fake legal cases used as supporting evidence in a new defense, or the production of seemingly valid numerical facts that are totally made up. All these possibilities can be generated by models, and they are especially dangerous when the task at hand is to use underlying supporting data as evidence to draw a conclusion. This is true particularly in data analytics, where there isn't one defining fact or figure that can accurately capture the nuance and complexity of reality.

You can see model hallucination play out when you ask AI to explain how it arrived at an answer. Since the whole model is opaque technology that provides predictions, the sound reasoning that you may be familiar with receiving from a human is typically not available. You can also see model hallucinations and limitations when it outputs what you know to be a wrong answer and you try to coax the AI into arriving at the right answer. Figure 1-3 shows a conversation I recently had with ChatGPT about the popular TV show *Survivor* and the phrase its host uses at challenges. In this exchange, the model initially provides a factually incorrect answer, including the word "guys" at the end of the phrase. Only after redirecting the model to consider more information, like that it was updated in a recent season, does it return the factually accurate answer. This isn't hard to imagine, since there were 40 seasons of Survivor where "guys" was part of the phrasing, likely influencing the model more heavily to include the word "guys" and ignoring more recent history.

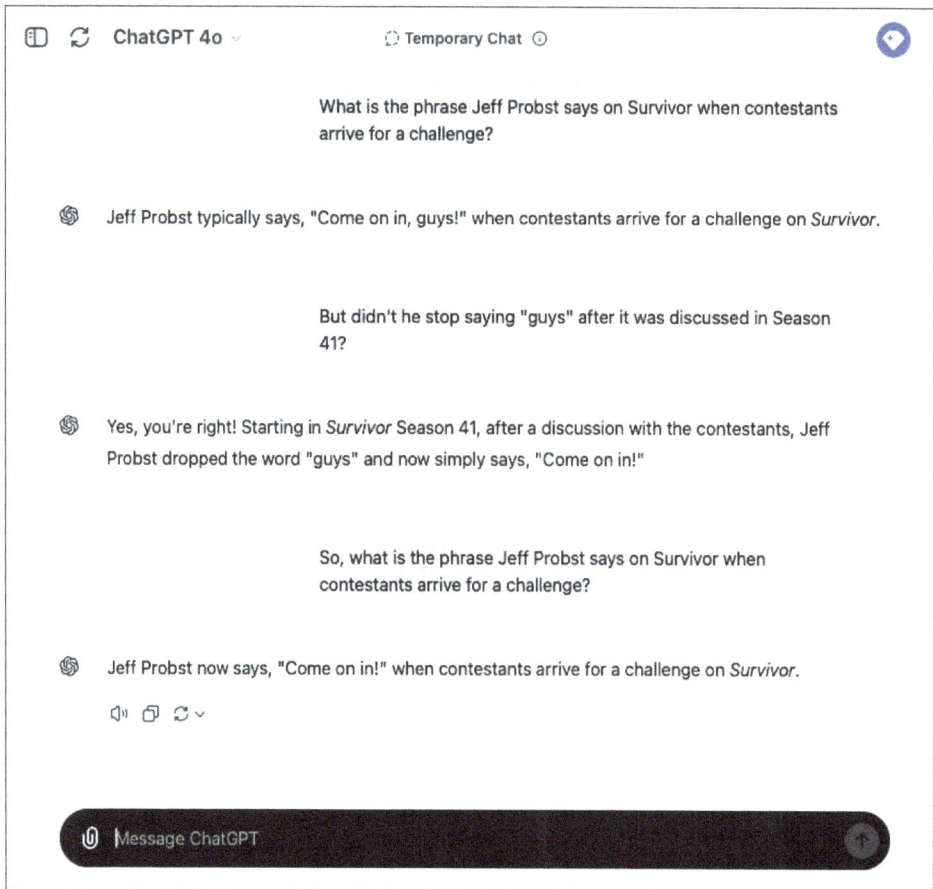

ChatGPT 4o ⌄   ⊖ Temporary Chat ⓘ

What is the phrase Jeff Probst says on Survivor when contestants arrive for a challenge?

Jeff Probst typically says, "Come on in, guys!" when contestants arrive for a challenge on *Survivor*.

But didn't he stop saying "guys" after it was discussed in Season 41?

Yes, you're right! Starting in *Survivor* Season 41, after a discussion with the contestants, Jeff Probst dropped the word "guys" and now simply says, "Come on in!"

So, what is the phrase Jeff Probst says on Survivor when contestants arrive for a challenge?

Jeff Probst now says, "Come on in!" when contestants arrive for a challenge on *Survivor*.

Message ChatGPT

Figure 1-3. A discussion with ChatGPT about Jeff Probst

Since the hallucination phenomenon tends to be more prominent in ambiguous situations—when the question itself may be unclear, the training data is insufficient, or extrapolation is required—it's crucial to remain skeptical. These scenarios are common in the world of analytics, where understanding the reasoning behind insights is essential. Historically, human experience and intuition have been key to overcoming these challenges—an ability AI lacks, making it harder to fully trust its conclusions without deeper scrutiny.

Worker Displacement

AI holds the promise, danger, and fear of displacing humans who are doing the work. In data analytics, there is a very real sense that if an end user can interact directly with AI to get the analytics and insights they need, the utility of the analytics professional is displaced. It can be frightening to consider what your actual job is if the AI can—with a few simple text commands—do tasks like building charts, curating data sets, or constructing complex queries that have been your full-time job. One of AI's major marketing pitches is that AI will improve efficiency, but *improved efficiency* almost always means that someone will be made redundant in the process.

Although the prospect of AI taking away jobs may seem scary and inevitable, this is an area where data professionals must be loud advocates and have confidence in the body of work they have produced. The charge of data analytics hasn't changed. The analyst's responsibility and purpose have always been to ensure that the analytics being produced can be trusted and is useful. If anything, that purpose becomes much more necessary when computers start doing the work unsupervised.

The tasks of data analytics can and will change with this shift. But remember that new tasks will usher in new jobs and responsibilities, much like the story I shared with you in the Preface of how improvements to Tableau Desktop's connectivity to multiple databases changed my requests from standalone data sets to requests for direct access to databases. It is still the responsibility of the data professional to manage the analytics, manage the access, manage the information produced, and manage the data products that are available to audiences. And in an applied situation, such as when Tableau Agent starts producing charts automatically, the analyst must still ensure that the chart constructed is accurate. The learned human skills and knowledge to make that discernment don't disappear over time; they become even more mission critical.

Finally, a reverse force is working in humans' favor. One of the most often-heard soundbites in data analytics is how many people are *underserved* with data and analytics. A disparity exists between the massive amounts of data being collected and the availability and comprehension of said data to workers. So if anything, it would be sound to reason that the audiences for data analytics are bound to grow as tools for easier access become available to them. And if you're a true zealot like me, obsessed with democratizing availability of data for decision making, it is an exciting time to be the sherpa of this practice to a broader group of people.

Trusting AI

This discussion of the risks associated with LLMs and generative AI leads naturally to a discussion of how Tableau and Salesforce handle trust and risk mitigation. The primary arbiter of trust with these technologies is called the *Einstein Trust Layer*. You can consider this layer as an intermediary between your enterprise data and the LLM models (at the time of writing, Tableau uses OpenAI's GPT-3.5 Turbo) that generate summaries and respond to your prompts/queries. Figure 1-4 shows the relationships among Tableau Cloud, the Einstein Trust Layer, and LLM models.

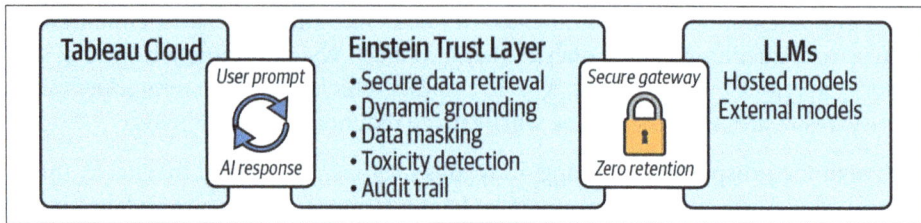

Tableau Cloud	Einstein Trust Layer	LLMs
User prompt ↻ *AI response*	• Secure data retrieval • Dynamic grounding • Data masking • Toxicity detection • Audit trail	*Secure gateway* 🔒 *Zero retention*	Hosted models External models

Figure 1-4. The relationships among Tableau Cloud, Einstein Trust Layer, and LLMs

The key components of the Einstein Trust Layer include the following:

Secure data retrieval
> Access controls and permissions are built into the Tableau platform. These controls are managed by administrators and creators and dictate who has access to data sources, dashboards, and Pulse metrics. During the process of interacting with AI, these controls are checked and verified prior to generating a response or serving up an insight.

Dynamic grounding
> The LLMs are enriched with context related to your business. Additionally, when Pulse metrics are served up, templates are used as guidelines for the prompt response. This process includes very clear instructions to the LLM model on not guessing, using neutral language, and handling the numerical values it receives. It also gives clear instructions on the format of the response, like the inclusion of time period comparisons.

Data masking
> To eliminate passing of protected information, like *personally identifiable information* (PII), sensitive values or words (such as a customer's name) are *tokenized*. In the case of a customer's name, the name isn't sent to the model, but instead a substitute token or string of text is used. Upon serving up the result, the token is then translated back to the PII it represented.

Toxicity detection
> This is the process of scoring prompt results or generated summaries for harmful information, which includes language that exhibits hate, violence, identity, or sexual content. Each response is scored, and if it is overly toxic, it will not be served back to the end user.

Auditing
> An audit trail is created and stored in Salesforce's Data Cloud associated with the prompt or information processed by the LLM. This includes the user who initiated the prompt, the body of the prompt (both masked and unmasked), a categorization of whether PII was found in the prompt, the toxicity score, and any user feedback associated with the output they receive.

Zero retention

After the prompt is resolved, the prompt itself is purged from the system. This ensures that the LLM doesn't retain the prompt as future information it has "learned" to influence the next response or the overall model.

> Beyond the Einstein Trust Layer, all data is encrypted both at rest and in transit. Salesforce uses OpenAI's GPT models and has agreements in place that dictate OpenAI will not store or retain any information or prompts that are fed into the models. And if you're deeply curious, all these services are hosted on Amazon Web Services (AWS), which exists globally across many availability regions. For a deeper dive or explanation, watch this Salesforce presentation (*https://oreil.ly/zExox*) on the topic.

Competitors and AI

As this chapter comes to a close, I want to touch on how competitors to Tableau are imagining and implementing AI into their business intelligence (BI) platforms. While many options certainly are available for comparison, I've chosen to focus on Microsoft Power BI (PBI) and Google Looker.

AI in Power BI

Microsoft Power BI has three distinct types of AI features:

Advanced analytics assisted by AI

This feature includes baked-in advanced analysis types available to those working in PBI without the need to code. Of note is anomaly detection, which scans data presented in a visualization or data set and provides statistical results aimed at finding anything out of the ordinary. Two advanced data exploration options, take the form of interactive charts. One is called *key influencers*, an AI-driven visual aimed at finding primary drivers, and the other is a decomposition tree. The *decomposition tree* breaks a metric into the (likely) hierarchical dimensions and categories that exist within a data set. Sentiment analysis and forecasting are also available as no-code solutions.

AI assistance for analytics creation

Very similar to Tableau Agent, this is an interactive AI chatbot that can help developers create Data Analysis Expressions (DAX) code, a formula expression language. There are also areas throughout Microsoft Office products, like Teams and SharePoint, which can take suggested starting visualizations or analytical questions and automatically build them in PBI.

Natural language query (NLQ)
This is nearly identical to the concept of Ask Data or interacting with Tableau Pulse. Here users can formulate questions that are turned into queries and return visualizations and results. The distinction to remember is that the input is coming from a nontechnical business user.

AI in Looker

Google's listed and upcoming AI capabilities nearly mirror what I've already described for Tableau and Power BI. Of note, they are focusing on the following:

Duet AI assistant
An AI assistant designed to help analytics creators make visualizations and access data. Notably, this includes methods to translate natural language requests into its own proprietary query language, LookML.

AI-powered visualizations
Again, this is technology designed for a business user to automatically create visualizations based on information and data they have access to.

Summarized insights
While there aren't many specifics on how or where this will be implemented, this is fundamentally similar to the insight summaries served up in Tableau Pulse.

Of note in discussing these competitors are the AI models that are the engines. Microsoft has a large stake in OpenAI and utilizes that company's models or variations of its models. Google, on the other hand, leans on its own proprietary LLMs—namely, Gemini (formerly Bard).

Summary

In this chapter, you've learned about AI in the analytics space. Key applications of AI in analytics include the following:

- More sophisticated analysis methods, like predictive multivariate models and sentiment analysis (scoring of text as positive/negative/neutral)
- Natural language processing and querying (NLP and NLQ), which allow users to ask questions in normal (human) language and receive analytical results
- Generative AI that summarizes information and surfaces interesting insights to end users
- AI assistance that makes building analytics easier

To help you understand the latest AI tools in this book, the chapter unpacked how LLMs function:

- They consume large amounts of text and categorize words across a copious number of dimensions.
- Prompts and text inputs are processed in layers called *transformers*. These models tend to first resolve sentence structure and ultimately reach comprehension and context of the input.
- Humans control what training data is fed into the models as well as guidelines for processing that information. However, much of what occurs within the LLM is unknown and left to the model.

It is important to remember that LLMs and generative AI are essentially very sophisticated prediction engines. They rely heavily on the training data they receive. The output is influenced in how the LLM was constructed and whatever instructions it was given. Additionally, there are some inherent risks to be aware of when working with generative AI:

Model bias
An LLM can have bias built in from the training data it receives. The model relies on information that has already been recorded by humans. In particular, there can be biases around societal issues (like diversity), and responses can be heavily influenced by user input.

Model hallucination
An LLM could respond with a factually made up or erroneous result. This typically occurs in ambiguous situations or where extrapolation is involved—both of which occur frequently in analytics.

Worker displacement
Although it can cause worry to consider how analytics roles may change, the bedrock of analytical skills you hold are necessary to ensure that the AI-derived results are grounded in sound analytical reasoning.

Tableau and Salesforce aim at creating trust while using AI-powered tools via the Einstein Trust Layer. This layer acts as a go-between for enterprise data and the LLM engines processing prompts. The six facets of this trust layer are as follows:

- Secure data retrieval
- Dynamic grounding
- Data masking
- Toxicity detection
- Auditing
- Zero retention

And finally, two major competitors of Tableau, Microsoft Power BI and Google Looker, show common themes in the utility of AI in the analytics space, namely: AI assistance in analytics creation, advanced mathematical analyses, summarized insights, and natural language query.

In Chapter 2, I'll show you how to get started with Tableau Pulse. I'll break down how to activate it in your Tableau environment, how to build your first metric, and more. By the end of the chapter, you'll have the confidence and knowledge to start deploying your own metrics for business users to start following. And with the history of Tableau and knowledge of LLMs you received here, you will be able to accurately convey AI's role and the risks when using it in Tableau.

CHAPTER 2

Getting Started with Tableau Pulse

Now that you have a history lesson on Tableau and AI in analytics under your belt, it's time to dive in to getting hands-on experience with Tableau Pulse. In this chapter, I'll take you step by step through everything you'll need to know to get your first Pulse metric ready. I'll walk you through key concepts and definitions, and you'll see detailed images to help you navigate the platform.

Purpose and Key Definitions

I recently saw a presentation on Tableau Pulse about its purpose. The presenter said that Pulse was designed to answer the 20% of questions that users ask 80% of the time. This is a common trope in data analytics, the Pareto principle, that helps to give an understanding of how Pulse fits into the Tableau ecosystem. To unpack this 80/20 principle a bit further, the speaker posited that end users tend to ask the same small subset of questions (the 20%) a majority of the time (80%). As an analytics professional, you might find this a little appalling, but take a step back, and there's a decent amount of truth there. Imagine a social media manager who is tracking a campaign on Instagram. She may want to start her day by looking at key metrics that give her a sense of how the campaign is performing. Similarly, when you first get into your car, you may check the fuel gauge, see if the check engine light is on, and monitor tire pressure. That's how you can imagine Pulse, and while I've not found any documentation to confirm, I have to imagine its name is based on the phrase *keeping your finger on the pulse.*

Prerequisites

As mentioned in the Preface, Pulse is currently available only within Tableau Cloud. Tableau Cloud is the software as a service (SaaS) solution of Tableau Server, the web application that organizes, stores, and serves up Tableau content. Based on the architecture of how Pulse functions (see Figure 2-1), I imagine that this limitation is due to the necessity for information to be taken from Tableau and passed through to an LLM that is completely separate and unrelated to the Tableau software. Similarly, the audit trail of the prompts would necessitate that information gets written to the Salesforce Data Cloud, which would require that the Tableau environment is available on the internet (not just a corporate intranet).

Pulse is disabled on each Tableau Cloud site by default. To enable Pulse, navigate to the site settings. In the General section, you'll be given the option to turn it on for all users within a site or a single specified group of users. This configuration section includes some info on what Pulse does and explains that it is currently being updated on a frequent (biweekly) basis. Figure 2-1 shows where to enable Pulse.

> To have access to the Settings section in Tableau Cloud, you must have site administrator permissions. This means your license role must be either a Site Administrator Creator or Site Administrator Explorer.

In addition to enabling Pulse on your site, you must have at least one published data source to generate a metric. A *published data source* is a standalone object in Tableau that represents a data set for a specific analysis. Inclusive of a published data source are the following components:

Data connection
> The credentials, connection information, and relationship (joins) between data tables used in the data set. This information is all packaged up so that each subsequent user doesn't have to redo this work.

Field customizations and defaults
> Fields in the data set can be refined to have friendly field names (as opposed to the names in their source systems), number format, default colors, default aggregations, and definitions. Unnecessary fields can be hidden from end users, yielding them unavailable for analysis. These all define the default behavior of how a field will be used when constructing visualizations.

Added fields for analysis

New calculations based on the underlying data can be constructed (imagine having Sales and Cost of Goods—Profit would be the mathematical difference between those two). Additionally, new dimensions and analytical tools like bins, sets, and parameters can be included.

Data freshness

Specifications as to whether the data source is connected live to the source system or extracted and refreshed on a specific schedule. An *extract* is a snapshot of the data from the source system, which can be given a frequency for the system to retrieve new/updated data.

Description and certification

These include a user-friendly description on what is included in the data set; typical descriptions include the source systems and how the data source can be used for analysis. A *certification* badge can also be applied, which links the username of the person who certified the data and an optional message with information about the certification status.

Permissions

Since it is a standalone object, the permissions for who can access, modify, and delete a published data source must be set.

The purpose of a published data source is to be a *single source of truth* for a particular scope of analysis. Its key benefit is the knowledge that each subsequent analysis, visualization, and metric are functioning off the same underlying data. Additionally, a published data source minimizes data-source proliferation and unnecessarily redundant connections to the same data in source systems. In a very mature Tableau environment, you can imagine that one published data source can fuel a whole body of analytical content (multiple dashboards, ad hoc analyses, and metrics).

> All Pulse metrics must be constructed from fields in a published data source. Data sources that are embedded within a workbook will not be accessible to create a metric.

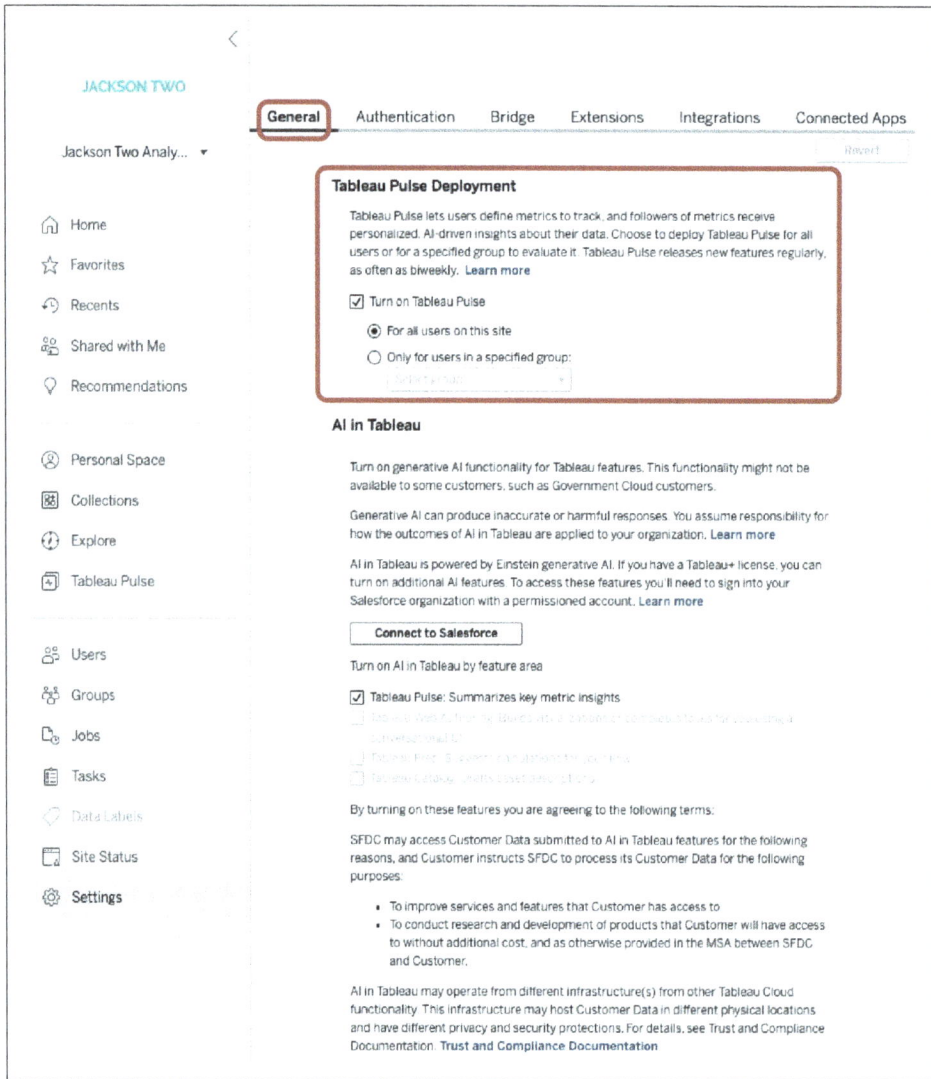

Figure 2-1. The General settings section within Tableau Cloud (see a larger version of this figure online (https://oreil.ly/lait0201))

Metric Definitions and Metrics

Up to this point, I've been calling the measurements available for following in Pulse *metrics*. While this is true, there are actually two subconcepts of metrics I'd like to unpack. First is the *metric definition*, which is the metadata specified for a metric and includes the following:

Name

The user-friendly name given to the metric, like *Sales*

Measure and aggregation

The numerical field being used to generate the metric and the aggregation (like sum/min/max) that is applied to it

Time dimension

Each metric needs a date or datetime dimension for time-bound comparisons, another field in the data set

Compared to

A configuration on the interval of time for time-bound comparisons, like prior year, month, or week

Adjustable metric filters

These are (likely categorical) dimension fields that can be added as filters to limit the metric results by end users

Number format

How the numerical value should be represented, like as currency or percentage

Value going up is

A specification as to whether an increase in the value is favorable or unfavorable

You, the analytics professional, will be responsible for specifying metric definitions. Depending on the size of your organization or the scope of your role, you will likely be creating the published data source that feeds the definition as well.

> All metric definitions must be unique in Tableau Pulse. If a user attempts to create an identical metric definition from the same data source, an error will appear when they try to save the definition. Within the error message is a link to the existing identical metric definition.

The second subconcept is a metric itself. This is a *visual representation* of the metric definition that can include the application of the filters and time comparisons that were instantiated in the definition. Say, for example, you construct a sales metric that allows an end user to filter by region and set the time range for display. If there are four regions in your data, you can imagine that there could be four iterations of the metric, one for each region. Similarly, there can be metrics for different time intervals chosen (like this year, this month).

Your First Pulse Metric

Let's create a metric from the beginning. I'll walk you through generating a published data source, metric definition, and subsequent metrics using the Sample - Superstore data set that comes packaged with Tableau. After I go through the creation process, I'll do a deep dive on the windows, menus, and configurations seen along the way.

What is the Sample - Superstore data set? It is sample data from a fictional retail store that sells a variety of products, mostly things you would find in an office. Each row of data within the Superstore data set represents a line item on a receipt. Imagine you purchased two binders, three mousepads, and one bottle of water. There would be three records (rows) for your transaction, one for each unique product, which include the quantity of each. You would also see the order number repeated for each of these three records, since they were all purchased within the same transaction.

Building a Published Data Source

To build a published data source, you must first connect to data. Published data sources can be made in Tableau Desktop, Tableau Prep Builder, and Tableau Cloud. Here's how you can construct a published data source in Tableau Desktop. First connect to the Excel file that includes the Sample - Superstore data. For default Tableau installations, this file can be found in subfolders within the My Tableau Repository folder found in your Documents folder. Figure 2-2 shows where it is located on my Mac.

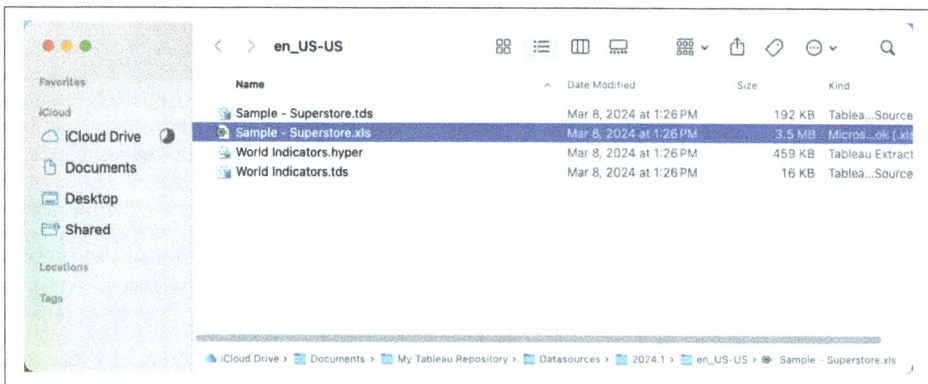

Figure 2-2. Location of Sample - Superstore data within My Tableau Repository

After selecting the *Sample - Superstore.xls* file, you're automatically taken to the Data Source screen to both preview the data and construct relationships among additional data sources or the tables contained within. Superstore has three tables associated with it: Orders, People, and Returns. For the sake of simplicity, you can skip relating tables and instead work exclusively with the Orders table. Figure 2-3 shows the Data Source screen after the Orders table has been dragged into the center.

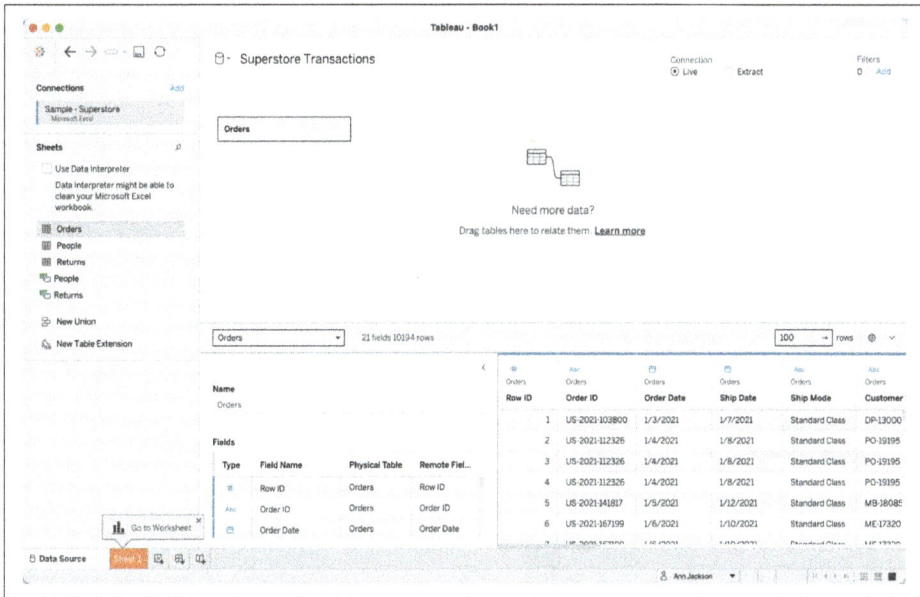

Figure 2-3. The Data Source area of Tableau Desktop with the Orders table in use (see a larger version of this figure online (https://oreil.ly/lait0203))

From here, navigate to Sheet 1. This will give you access to the Data pane, which is the area on the left side of the canvas that shows all the fields in the data set. The Data pane also includes a few prebuilt fields, denoted by italics, that Tableau includes for ease of use. You can see this in Figure 2-4.

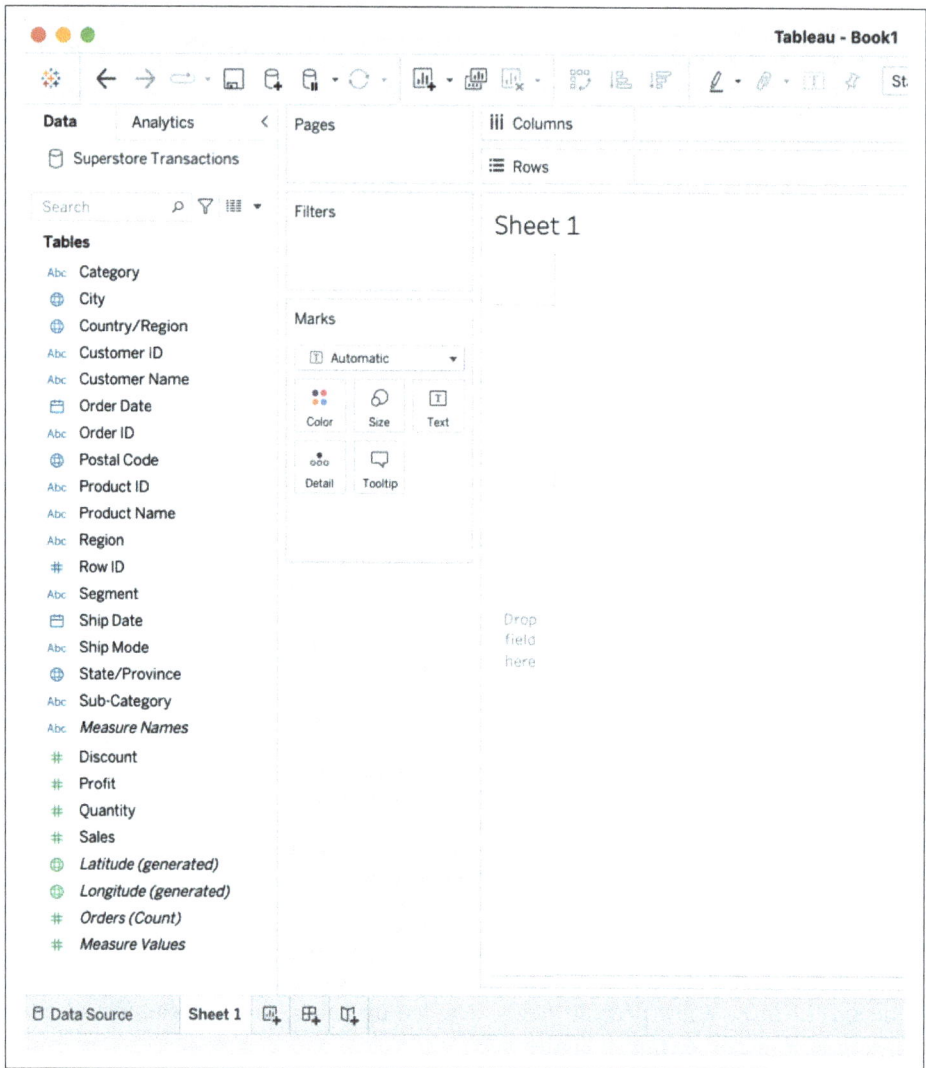

Figure 2-4. A blank worksheet with the fields available for analysis at left

At this point, you can begin preparing your data set for analysis. Remember, the goal of this process is to make a data source that can feed many types of analysis. You want to construct something that allows other users to quickly understand each of the fields and be able to use it for their own analyses. To assist in the process, you can make the following modifications and additions:

1. Default aggregation and number format for Discount. The Discount field represents a percentage value that was given as a discount off the sticker price of an item. Right-click this field and navigate to the Default Properties section to set the default Number Format for the field to Percentage with 2 decimals. Set the Aggregation to Average, since you don't want users summing up the percentage points of discount. You can see this menu in Figure 2-5.

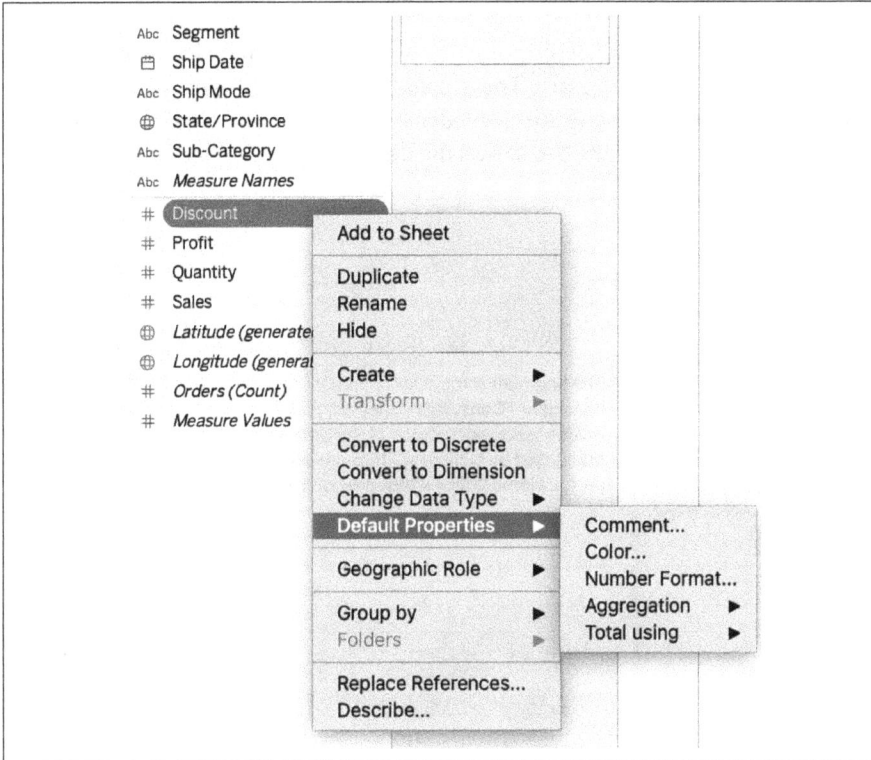

Figure 2-5. The Default Properties menu for the Discount field (see a larger version of this figure online (https://oreil.ly/lait0205))

2. Next, create a *hierarchy*, a way to organize dimensions into a parent-child relationship, for the geographic data included in the data set. To do this, drag one field on top of another in the Data pane and organize them from least granular geographic field (Country/Region) to most granular (Postal Code). You can also include the Region field, which is a value assigned by the organization.

3. Similar to defining the default properties of a field, you can assign the Region field a Geographic Role. This is done by selecting the "Create from" option and choosing the State/Province field.

4. Set the default number format for both Profit and Sales to be Currency (Standard).

5. Hide the Row ID field from use. This is an internal field generated by the Superstore database and isn't used in analysis.

6. Finally, you can add descriptions to a few fields—namely, Segment, Category, Order Date, and Ship Date. These fields are often confused within the organization, so adding a description (called a *comment*) will help guide anyone working with the data set. Figure 2-6 shows the description I've added for Segment.

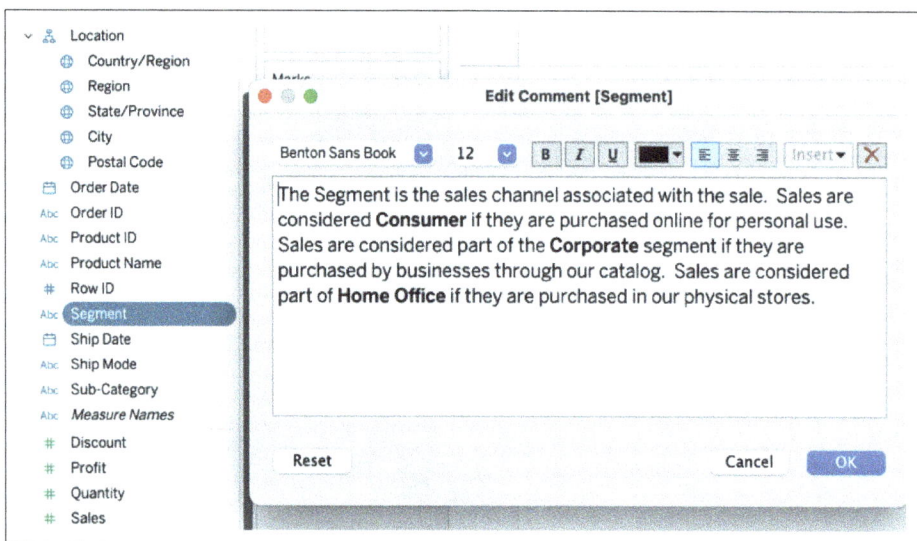

Figure 2-6. A description added to the Segment field

Now that you've customized and refined the data source, it's time to publish it up to Tableau Cloud. There are many ways to do this, but I'll walk you through my preferred method. Right-click the name of the Data Source in the Data pane. From here, select Publish to Server. A subsequent window pops up, shown in Figure 2-7, where you can customize the published data source.

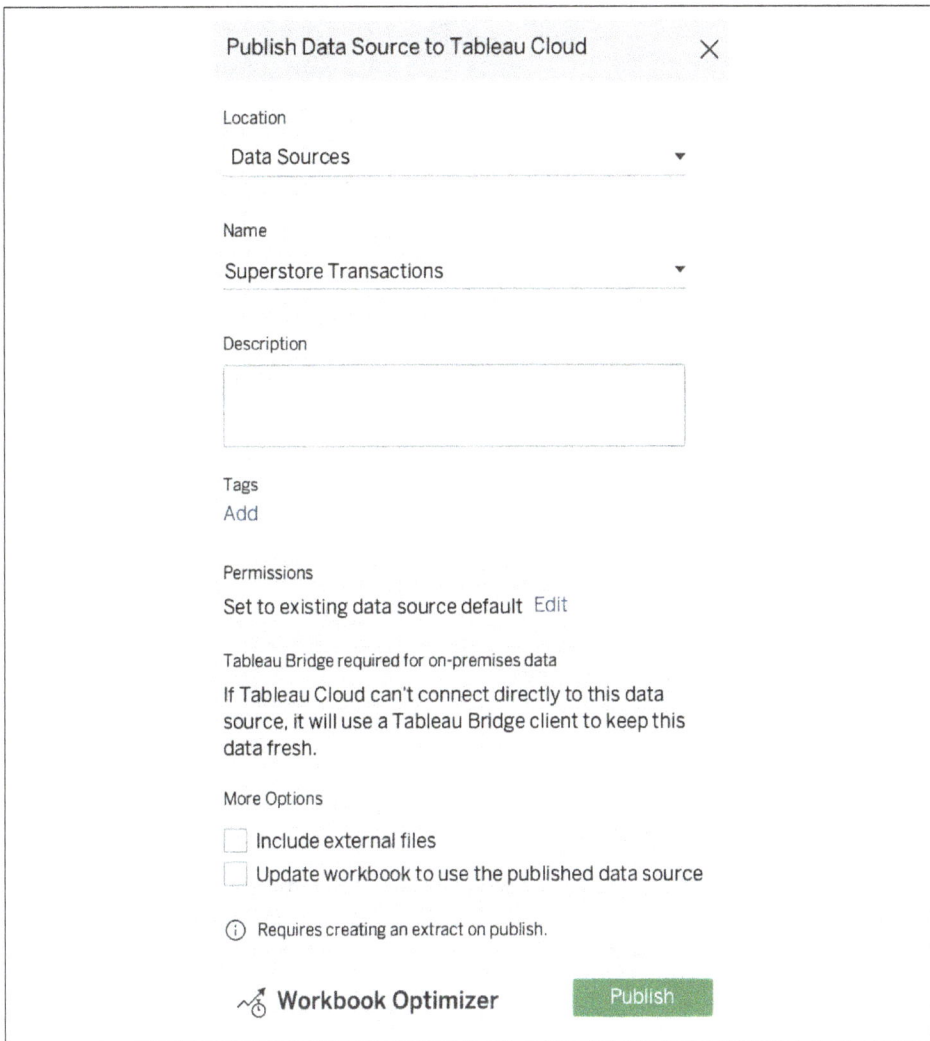

Figure 2-7. The Publish to Server window for a data source (see a larger version of this figure online (https://oreil.ly/lait0207))

Once the data source is published, a new window will open in your internet browser showing the location of the published data source in the Tableau Cloud environment, as shown in Figure 2-8. You can also certify the data source, modify the description, and add tags by clicking the information icon next to the data source's name.

You can now access this data source for building a Pulse metric definition. It can also be used by anyone with permissions to access it for their own workbooks or ad hoc analysis.

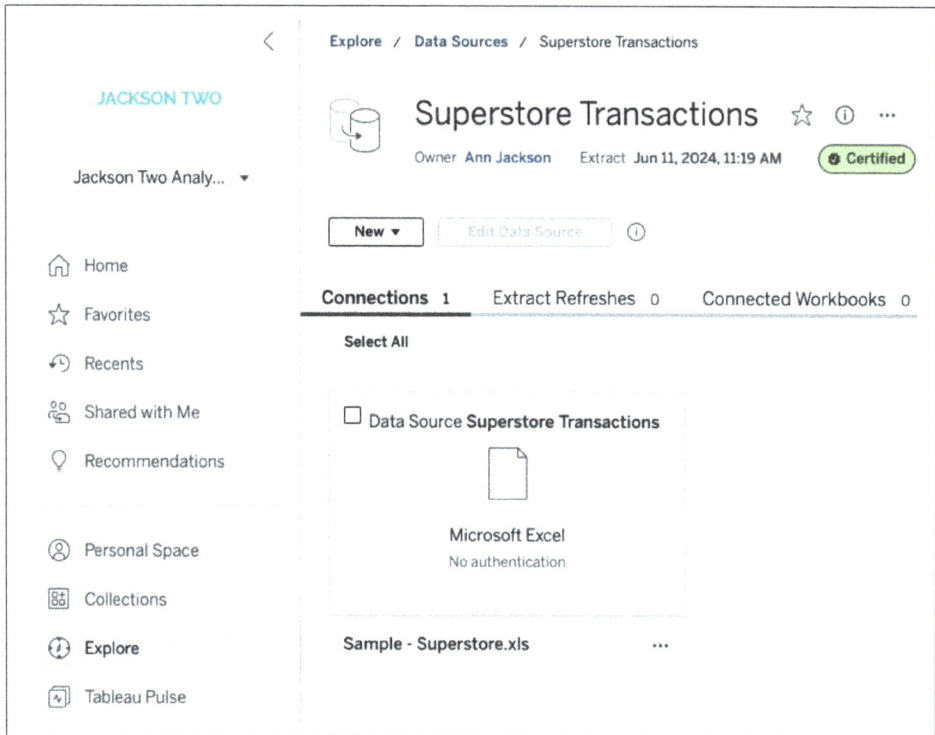

Figure 2-8. The final published data source

Building a Metric Definition

To begin creating a Pulse metric, click Tableau Pulse on the leftside navigation (beneath Explore), as seen in Figure 2-8. This will take you to a new screen, which at time of writing looks significantly different from the rest of the Tableau Cloud environment, as shown in Figure 2-9.

You can immediately click the New Metric Definition button and get started. Once it's clicked, a window will open to select the data source. Remember, this will include only data sources that have been published. The window includes a search function to easily find the data source you're looking for, and it includes key information about the data source itself. Figure 2-10 shows the selection window.

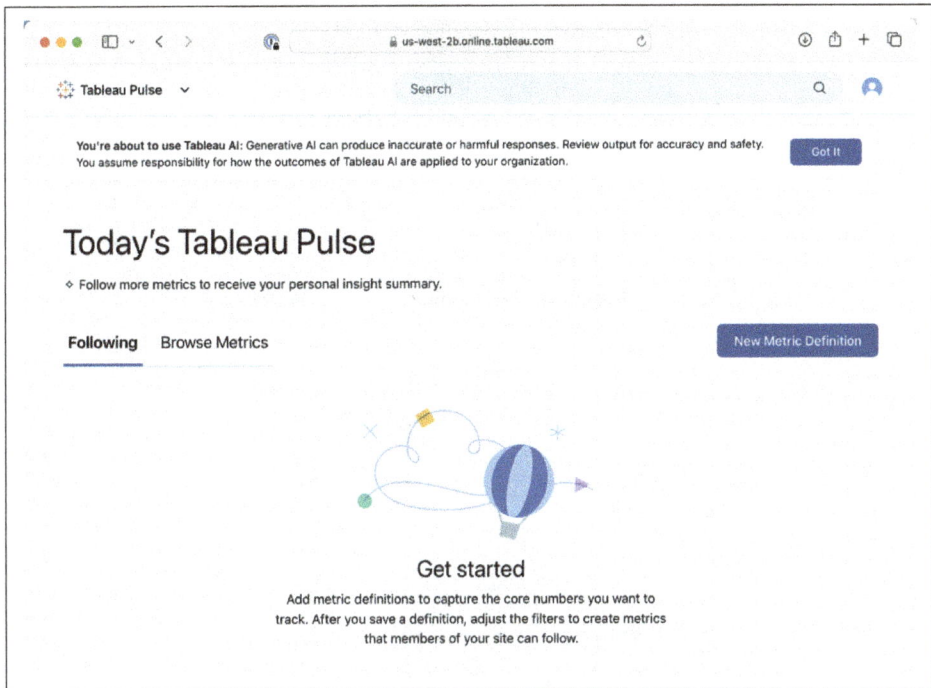

Figure 2-9. The Pulse section of Tableau Cloud

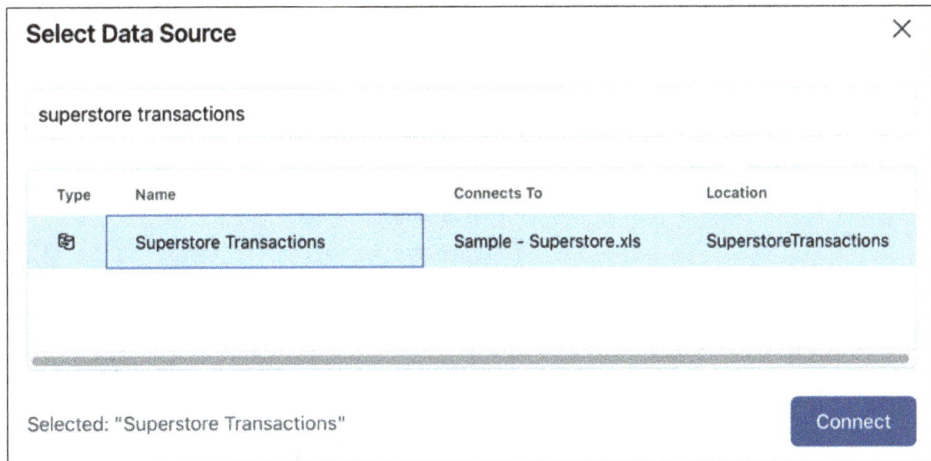

Figure 2-10. The Select Data Source window in Tableau Pulse

You're now at the heart of constructing your first metric definition and are presented with another window with a menu of configurable options on the left. The righthand side will change and update to include a preview of the metric as it is defined. Figure 2-11 shows the configuration options available.

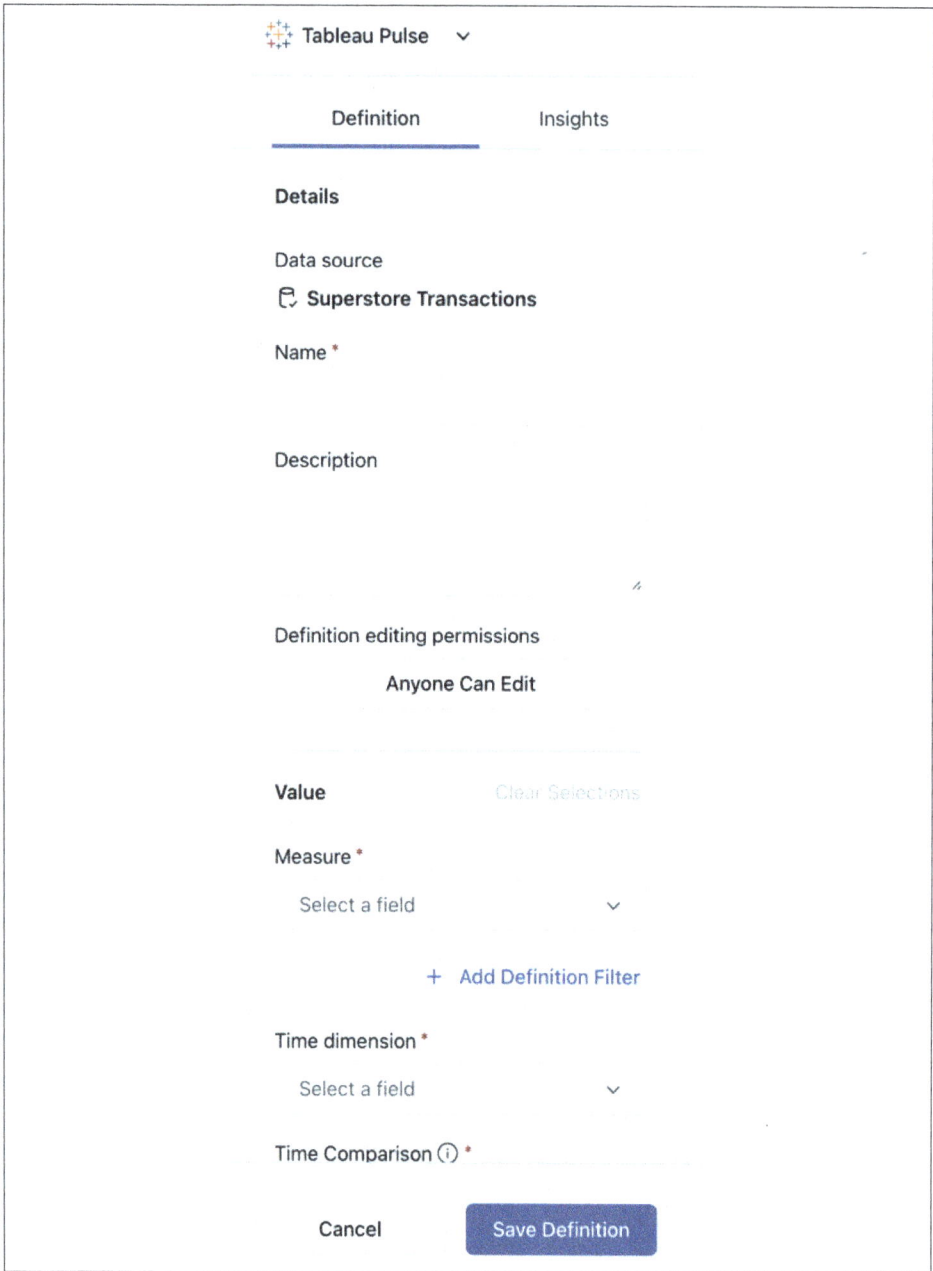

Figure 2-11. Metric definition configuration options

The first metric definition is going to be the Sales, or revenue, for Superstore, using the Order Date for trending and time comparisons. To immediately see a preview of the metric that Pulse will create, first specify the Measure as Sales and the Aggregation as Sum from the drop-down boxes. You must also specify a "Time dimension" (Order Date). Figure 2-12 shows a preview of what was created after specifying those configurations.

> If you're following along, your metric preview may look different from the images. Don't worry, this is expected, as Pulse relies heavily on time and will default to anchoring insights based on the date.

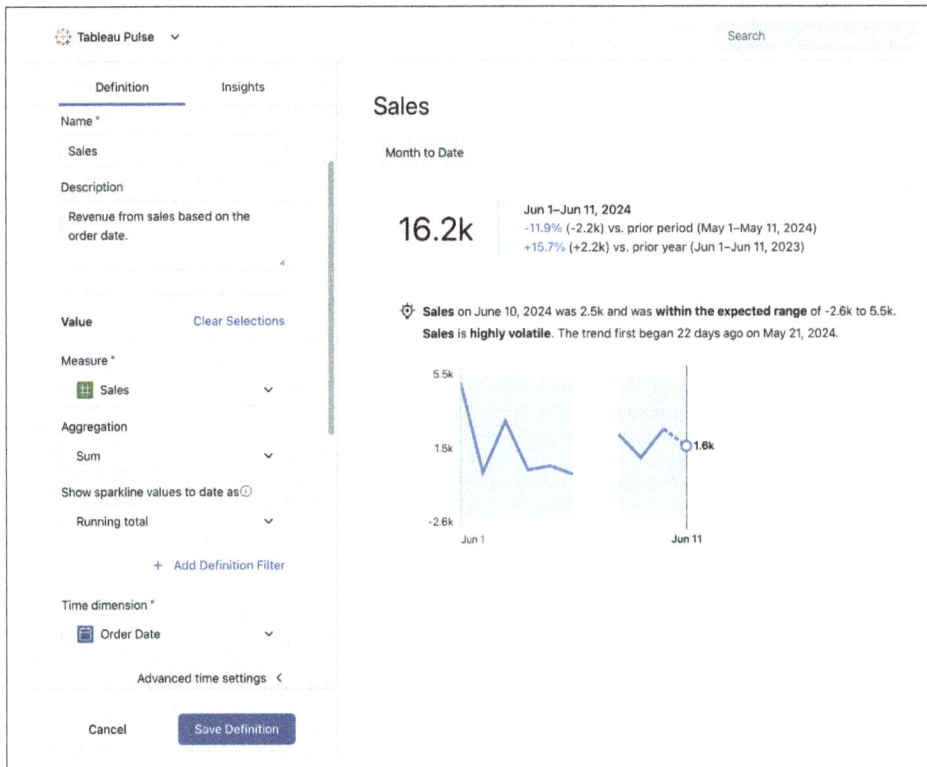

Figure 2-12. Preview of Pulse metric after selecting a Measure, Aggregation, and Time dimension (see a larger version of this figure online (https://oreil.ly/lait0212))

Continue down the Definition menu to the Options section. Here you can add "Adjustable metric filters," which are dimension fields that end users can select to filter the metric. Fields included as metric filters will also be used to determine insights for the metric. Add two fields to start: Segment and Category. Set the "Number format" to Currency to reinforce that the metric value is in dollars. The metric preview is updated with the two new filters near the top, next to the time period being displayed; the values now also show with dollar signs. Figure 2-13 shows the completed metric definition.

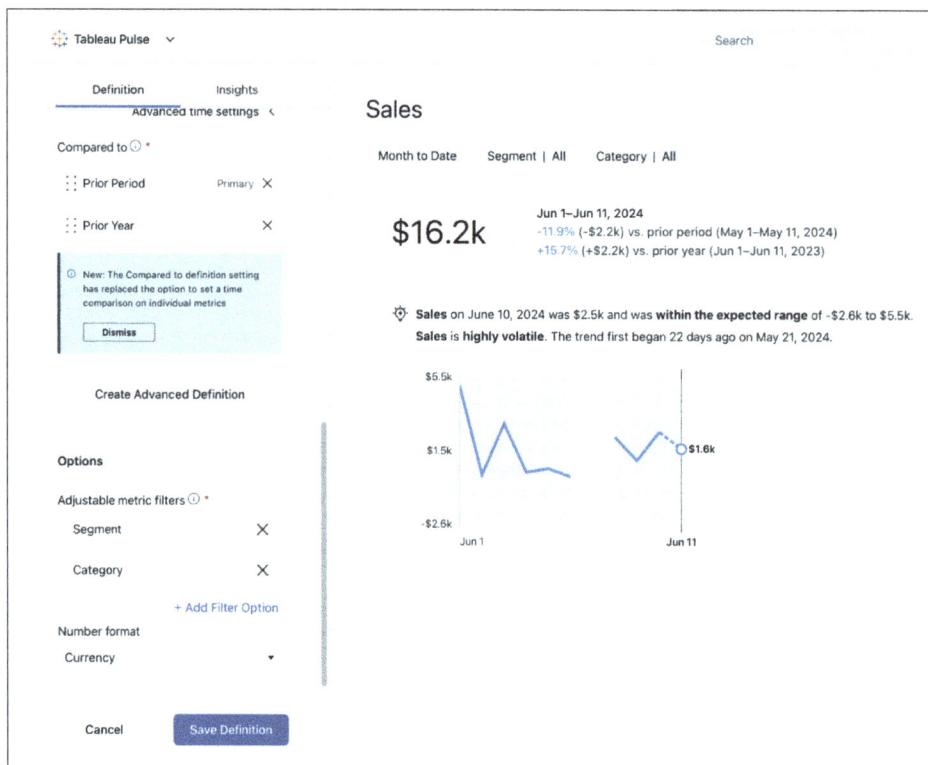

Figure 2-13. The completed metric definition with filters and formatting applied (see a larger version of this figure online (https://oreil.ly/lait0213))

After clicking Save Definition, the screen changes to show the newly constructed metric with generated data based on the metadata defined in the metric definition, as seen in Figure 2-14.

From this newly created metric, the end user can customize and save additional metrics to keep an eye on.

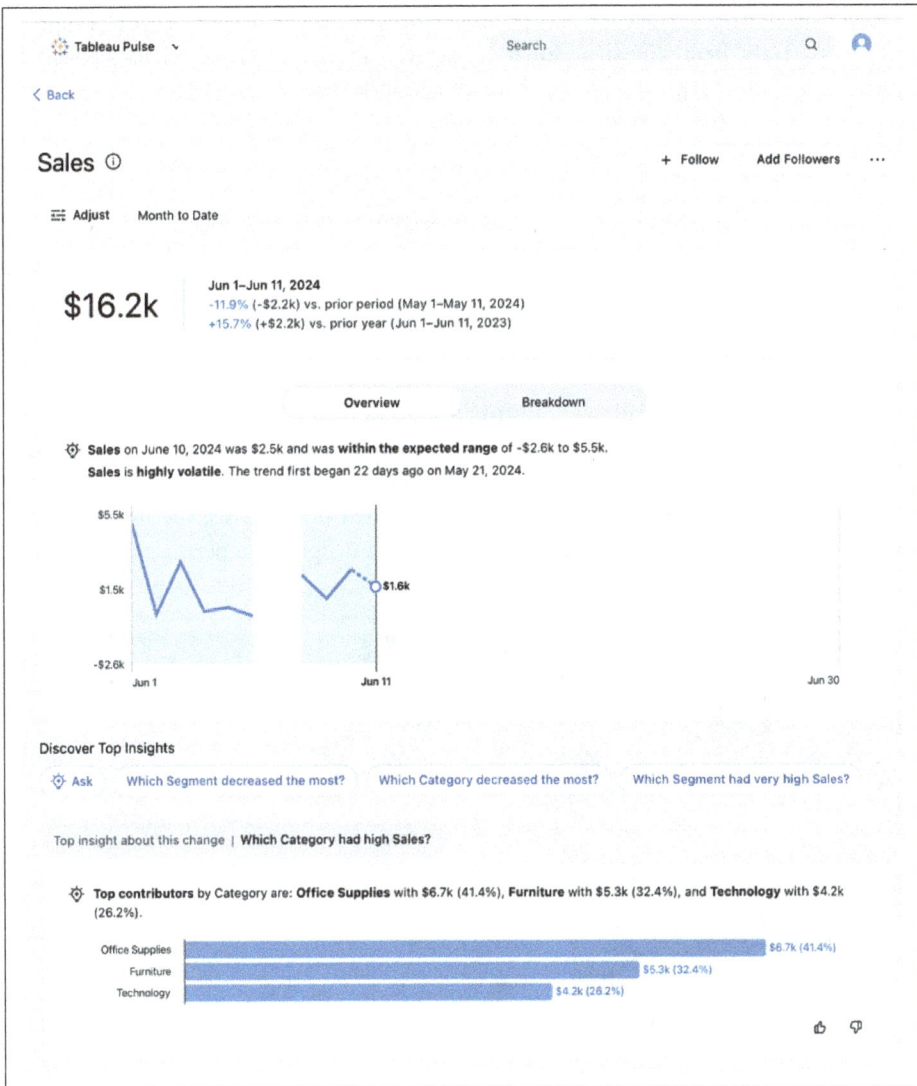

Figure 2-14. A metric based on the metric definition (see a larger version of this figure online (https://oreil.ly/lait0214))

Building an Additional Metric

Now you'll complete this example by constructing a metric for the "Director of the Corporate" segment. At Superstore, each sales channel is managed by independent verticals, so the "Director of Corporate Sales" isn't interested in the other two segments within Superstore. To do this, click the Adjust button directly below the name of the metric (see Figure 2-14) for options to select the time range of the metric and to filter the metric to just the Corporate segment. You can set the time range to Quarter to Date from the dropdown to show a longer trend of Sales, as seen in Figure 2-15. Don't forget to click the blue checkmark to save your selections.

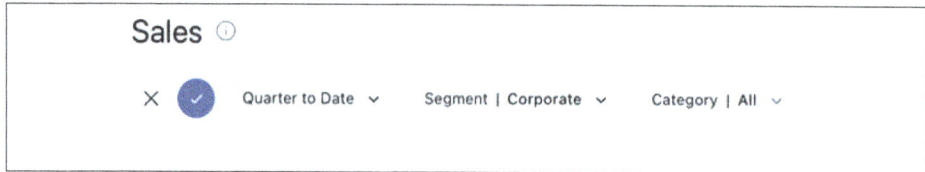

Figure 2-15. The options available for customizing the Sales metric within the Adjust menu

You can click the Add Followers button to select who should follow this metric. Figure 2-16 illustrates the selection of a user group (called Corporate Segment Users for the sake of the example). Once users are added as a follower, the metric will show up within their Pulse feed. You can click the Follow button to see the newly created metric in your own Pulse summary. Chapter 3 goes deeper into permissions for metrics and provides guidance on best practices.

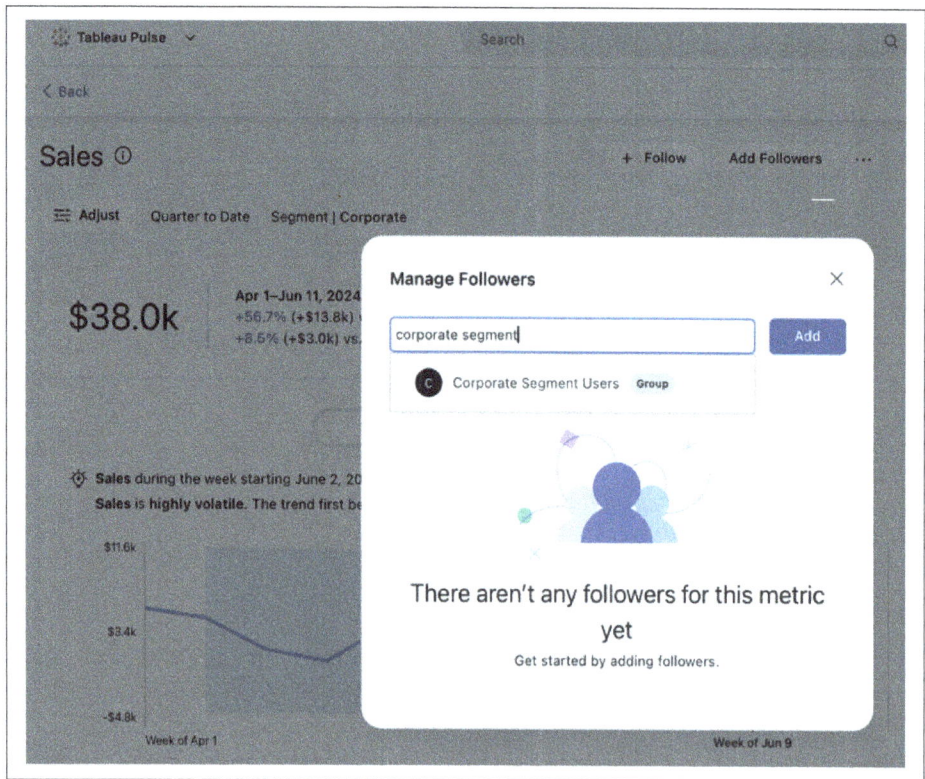

Figure 2-16. Selecting a group of followers for a metric

Accessing a Metric

Now that this custom Corporate segment metric has been created, return to the main area of Tableau Pulse. Since you follow the metric, it now shows up in the Following section, as seen in Figure 2-17.

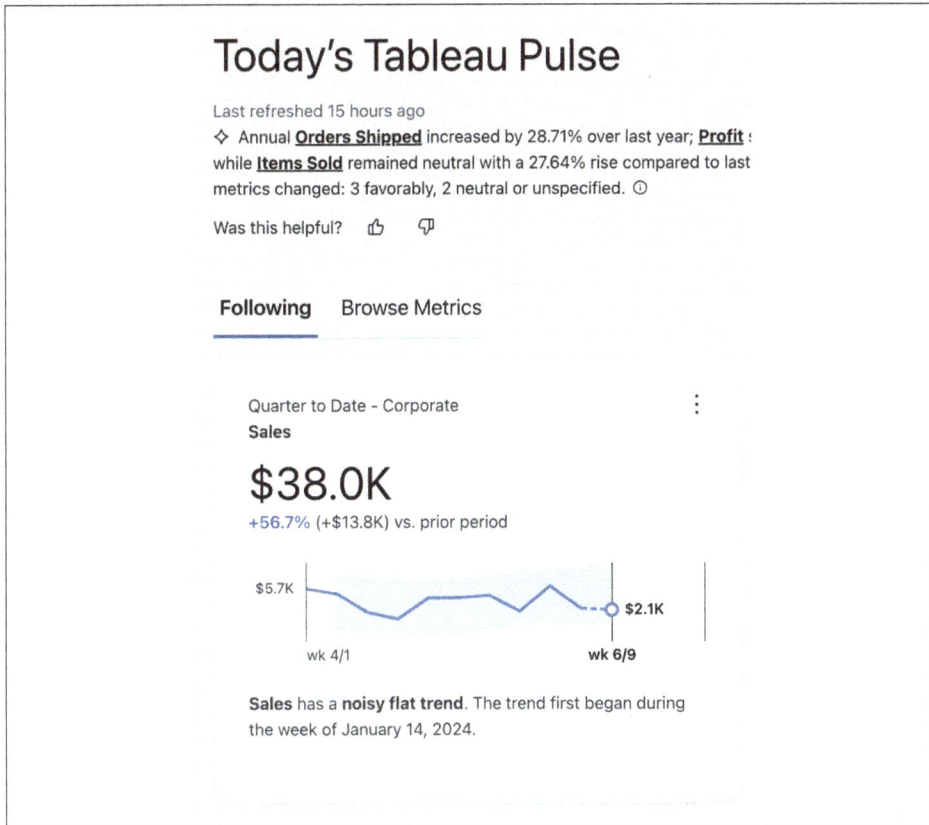

Figure 2-17. A Pulse summary with the Corporate segment sales metric (see a larger version of this figure online (https://oreil.ly/lait0217))

From this Following section, click the metric card to get back to the detailed view of the metric. Figure 2-18 shows the detailed view for the Corporate segment metric.

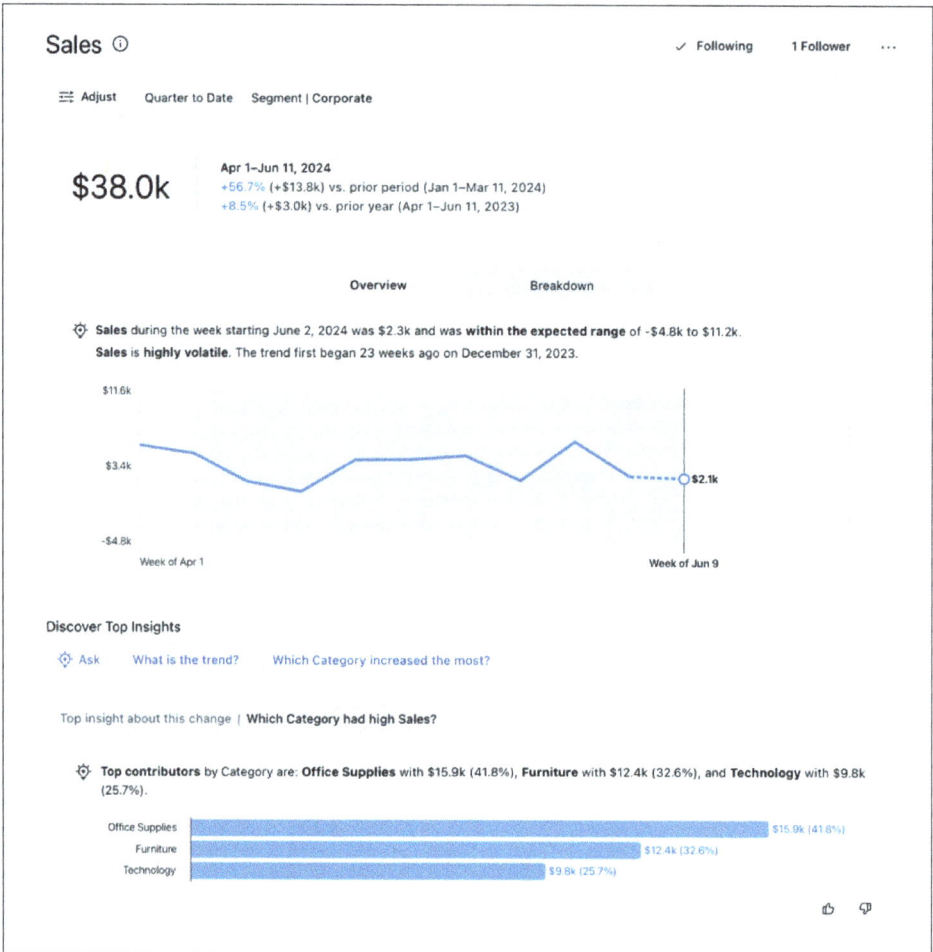

Figure 2-18. The Corporate segment metric detailed view (see a larger version of this figure online (https://oreil.ly/lait0218))

Interacting with a Metric

I'm now going to go through the sections within a metric that an end user can inter-act with. First, there are the Overview and Breakdown sections. The Overview sec-tion is exactly that, a visualization showing an aggregate of the metric for the time period, comparisons with other time periods, and a line chart of the sales (refer to Figure 2-18). If you click Breakdown, the visualization changes to show a bar chart that separates out Sales by each Category, as shown in Figure 2-19. This directly cor-responds with the adjustable metric filters configured in the metric definition.

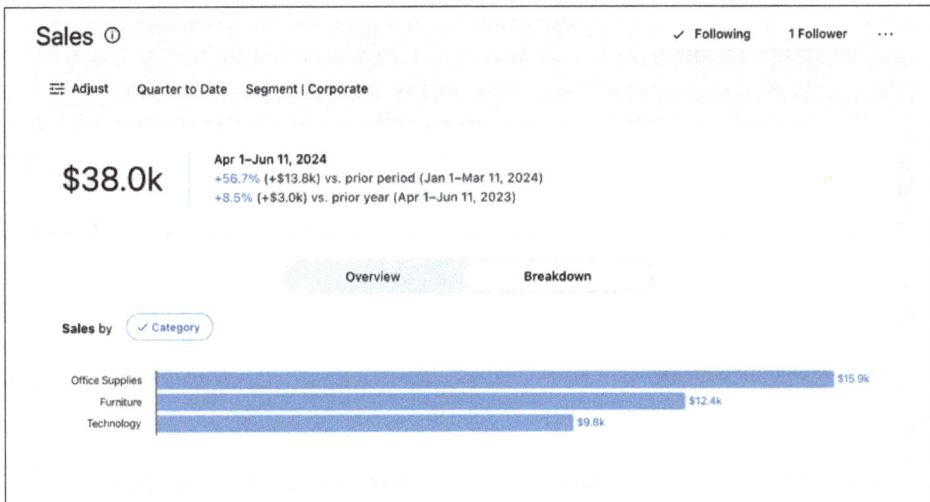

Figure 2-19. The Breakdown for the Corporate segment sales metric (see a larger version of this figure online (https://oreil.ly/lait0219))

In addition to this upper section, there is a bottom section labeled Discover Top Insights. Here you can see a few AI-generated questions that may be common or interesting for someone following the metric. There is also a visualization for the top insight associated with the metric. This metric shows a bar chart similar to the Breakdown chart, but with a summarized insights blurb and a percentage value for the distribution of sales among each category. You can click the thumbs up or thumbs down icons to provide feedback on whether the insight was helpful or not helpful to you, as seen in Figure 2-20.

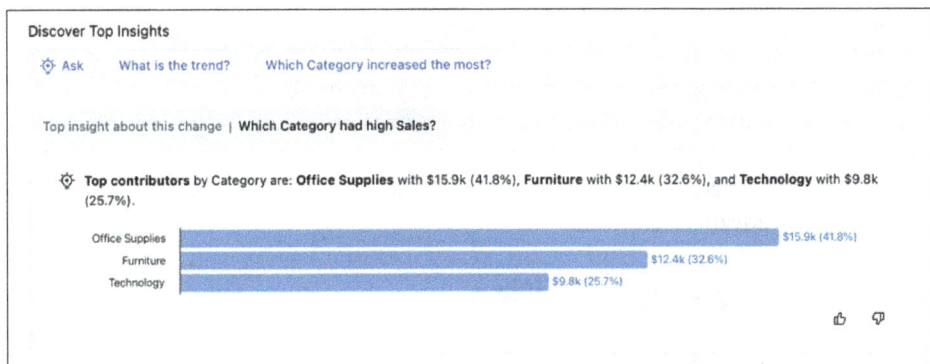

Figure 2-20. Top insights section associated with the Corporate segment sales metric (see a larger version of this figure online (https://oreil.ly/lait0220))

Clicking the other pregenerated questions shown (like "What is the trend?") causes additional visualizations with summaries to be displayed. Clicking the Ask button allows you to type in a question and see if an insight that may not be listed will surface. After entering your question, you'll receive either a new visualization or a list of suggested questions to ask if your question couldn't be resolved. Figure 2-21 shows what happens when asking for a trend of one specific Category, Office Supplies.

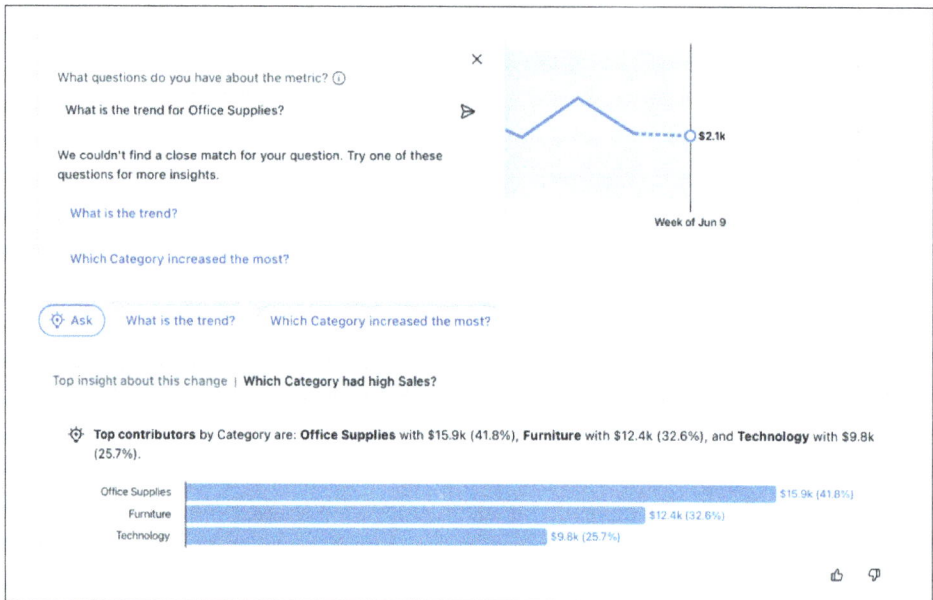

Figure 2-21. The Ask window interprets questions and suggests alternative questions (see a larger version of this figure online (https://oreil.ly/lait0221))

You've now seen a basic build of a published data source, metric definition, and metric. But as you may have noticed, more in-depth customizations and configurations can be made along the way. So the next section steps away from a specific metric and describes in more detail each of the configuration options available in the metric definition. Any configurations not covered in this chapter are discussed in Chapter 3, as those are more advanced features.

Dissecting a Metric Definition

You already saw what the metric definition configuration window looks like in Figure 2-11, but now I'm going to really unpack each section and venture over to the Insights tab of the window, which you may have noticed earlier. Going down the left-hand side of the Definition menu, you will see the following:

Data source

This is the name of the data source the metric definition is connected to.

Name

This is the name you give your metric definition. Ideally, this explains what the core definition is trying to measure. The red asterisk next to each configuration option denotes it is a required field.

Description

This is an optional field to include a description of what the metric definition is. This will propagate to every subsequent metric made from the definition.

When you constructed your first metric, you accessed most configurations in the Value section. Here's a full list that includes those and a few you haven't yet accessed:

Measure

Remember, this is the field that will be used for the definition. It can be any field type and isn't complete until an aggregation is specified for it.

Aggregation

Depending on the type of field chosen (namely, numerical or nonnumerical), you'll get different options. Table 2-1 shows a full breakdown.

Table 2-1. The aggregations available based on field type

Field type	Available aggregations
Number (integer and float)	Sum, Average, Median, Maximum, Minimum
Text (string) and Geographic	Count, Distinct Count
Date	Count, Distinct Count

Show sparkline values to date as

This is a configuration to determine if numerical values should be shown as a running total in certain visualizations or as non-cumulative. For Count and Distinct Count aggregations, this option is unavailable and set to non-cumulative.

Definition filters

These are filters that are applied to the metric definition and every subsequent metric. Unlike the Adjustable metric filters I constructed earlier, end users cannot interact with them, and they are preset. The best way to think about these filters is the default scope of the metric definition. For the previous example, you could include a filter for Country to show only Sales for the United States.

Time dimension

This is a required field that specifies which date field should be used for trending and the date axis in the metric visualizations.

Compared to

> These are prebuilt time-bound comparisons to be applied to the metric definition. They can be dragged and reordered to specify the priority and order in which they are displayed in the metric.

Create Advanced Definition

> This opens up a new window to define a metric based on a more sophisticated set of requirements. Chapter 3 covers this in detail.

Finally comes the Options section with the following:

Adjustable metric filters

> These are fields you can add to the metric definition that allow end users to use them as interactive filters. This is currently limited to text fields.

Number format

> This allows you to specify the formatting associated with the metric. Table 2-2 shows the options and behavior.

Table 2-2. How Tableau Pulse displays number formats for metric definitions

Number format	Formatting options and behavior
Number	Singular and plural units are appended after the numerical value (example: 1 widget, 2 widgets).
Currency	No options, automatically includes a currency symbol and displays the numbers with decimals. For large numbers, it condenses the number to a decimal with an appended thousands (k), millions (M), and billions (B) indicator, as appropriate.
Percentage	Returns decimal numbers as percentages with a % icon appended to the value.

As you saw in the example metric earlier, a metric definition can be constructed through the configuration of these fields. In addition, you can navigate over to the Insights tab to specify the types of information that are important for Pulse to surface within the metrics.

The Insights window, shown in Figure 2-22, has three sections:

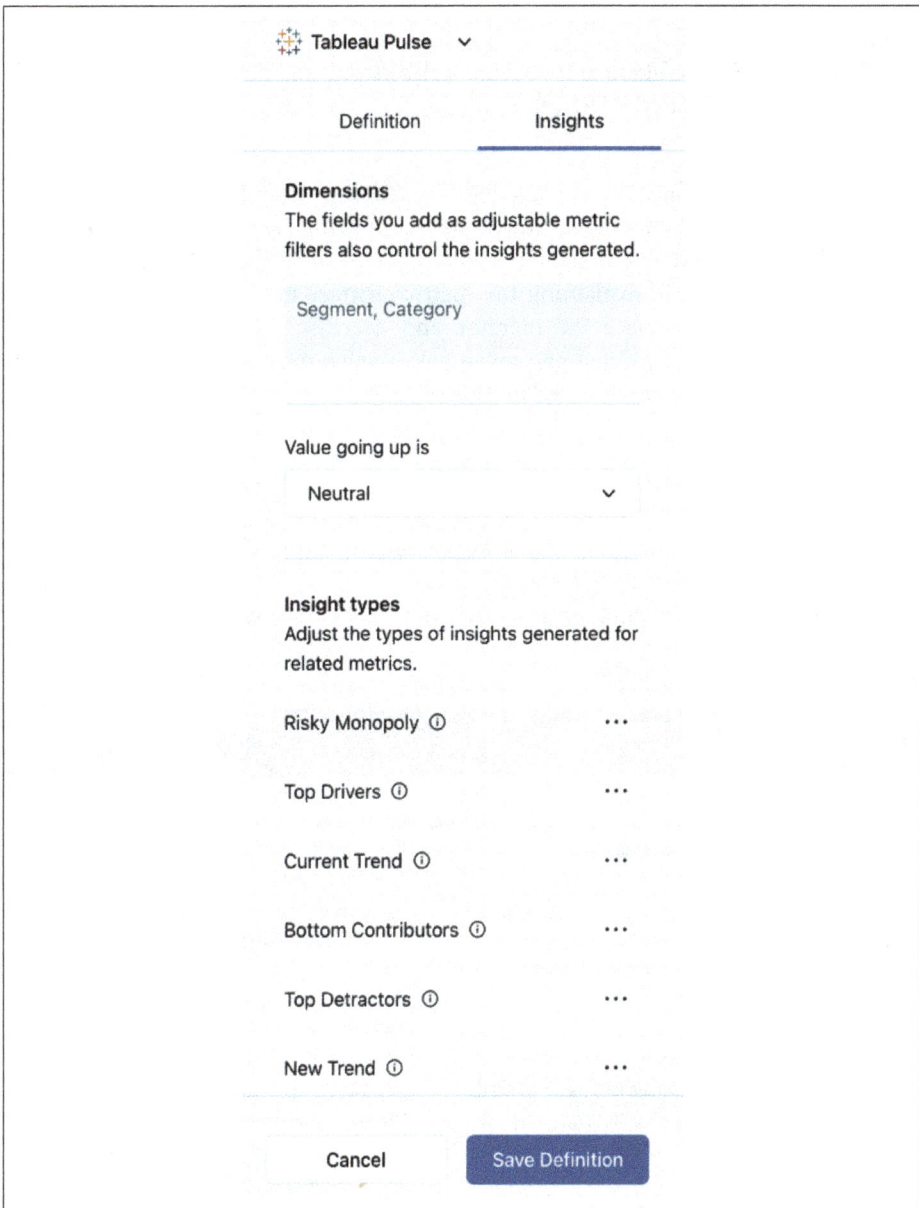

Figure 2-22. The Insights tab of a metric definition

Dimensions

This section is uneditable but automatically lists all the fields that were added as adjustable metrics in the Definition menu.

Value going up is

For the metric definition, you can define whether a numerical increase is Favorable, Neutral, or Unfavorable. These change the language used in the summary insights generated. Neutral is selected by default, which uses increase and decrease language in explaining the metric changes in the summary. Favorable and Unfavorable replace the increase and decrease based on what going up behavior is specified as. Red and green colors will also show up as additional visual encoding in the charts based on this selection.

Insight types

These are the ways in which the metric definition can be scanned and they determine the type of insights it will surface. By default, all types are turned on, but can be turned off using the triple dot actions menu to the right of the insight type. A handful of insight types are always on. Table 2-3 has a full listing with descriptions of the insight types and whether they can be turned on and off.

Table 2-3. Insight types used in Tableau Pulse

Insight type	Toggle use	Summary purpose	Behavior explained
Record-level Outliers	Yes	Shows row-level information for records that have an extremely high or low value for the metric.	When triggered, a table including the identified records will be displayed in the Insights section. This insight is useful for identifying bad data values or detecting anomalies.
Period Over Period Change	No	Shows how a metric has changed between two time periods. This insight is displayed as part of every metric.	This is a default part of insight summaries and compares the current metric value against the prior period. This insight helps quantify and track changes in a metric over time.
Top Contributors	No	Shows the highest values in a dimension for a metric within the metric's time range.	This is a default part of the metric breakdown. It shows values for the dimension members ranked at the top in terms of contribution to a metric's value. This insight helps users focus on high-impact areas driving a metric's performance.
Bottom Contributors	Yes	Shows the lowest values in a dimension for a metric within the metric's time range.	When triggered, shows values for dimension members ranked at the bottom in terms of contribution to a metric. This insight is useful when trying to identify areas of opportunity for improving a metric.

Insight type	Toggle use	Summary purpose	Behavior explained
Concentrated Contribution Alert (Risky Monopoly)	Yes	Shows when a small number of dimension members make up 50% or more of the contribution to a metric.	When triggered, shows the dimension members that contribute to the majority of a metric's value. This insight is useful in identifying business risk associated with an overreliance on a small set of contributors.
Top Drivers	Yes	Shows which dimension values for a metric changed the most in the direction of the observed change in the metric.	When triggered, shows values for dimension members that changed the most in the direction of the observed change in the metric. This insight is useful when trying to identify positive or negative drivers that contributed most to the corresponding change in a metric.
Top Detractors	Yes	Shows which dimension values for a metric changed the most in the opposite direction of the observed change in the metric.	When triggered, shows values for dimension members that changed the most in the opposite direction as the observed change in the metric. This insight is useful when trying to identify positive or negative drivers that contributed the most in the opposite direction of a metric's change.
Unusual Change	No	Shows when the value of a metric for a given time range is higher or lower than the expected range based on historical observations.	This is a default part of the metric's insight summary. This insight helps alert users to unexpected changes in a metric that may impact performance.
Current Trend	Yes	Shows current trends of a metric by communicating the rate of change, direction, and fluctuations of the metric's value.	Describes the current observed trend of a metric. This insight helps users stay on top of ongoing developments and changes in a metric.
Trend Change Alert	Yes	Shows an alert when the observed trend is dramatically different from the existing (current) trend.	When triggered, notifies the user to a change in the observed trend of a metric. This is useful for quickly identifying an unexpected change in a metric's performance.

Besides the different insights that are active for a metric definition, Tableau Pulse employs a scoring system to rank and surface the most statistically impactful insights to end users in their summaries. This becomes more apparent when a user follows multiple metrics and receives summaries of all their followed metrics in addition to individual insights for metrics. Additionally, this scoring system is influenced by the thumbs up/down feedback from a user on the deeper insights that surface.

Summary

You've now learned everything necessary to create your first Pulse metric and have a thorough understanding of the configuration options available. Here are some key takeaways:

- A published data source is required for building a Pulse metric.
- A published data source is a standalone object in Tableau that holds key information about a data set for analysis. It includes source system connection information, credentials, default settings (like formatting and descriptions), and additional fields added for analysis.
- A metric definition includes the measure field name, aggregation, time dimension, filterable dimensions, and insights available for all subsequent metrics.
- A metric is the interactive visualization and summaries based on a metric definition.
- Users can customize metrics based on the available filterable dimensions and preferred time ranges. Each configuration is saved as an additional metric that can then be followed.
- Metrics can be followed by a user or a user group and show up in each user's Pulse Summary.
- Many insight types can be applied to a metric definition. Most of them can be turned on or off based on the creator's preferences.

Chapter 3 explores advanced options within Pulse in more detail. You'll learn how to add definition filters, make advanced changes to time, and create an advanced metric definition. It also unpacks how calculated fields can be used in metric definitions. Finally, you'll learn how to manage permissions and metrics in the Tableau platform.

Advanced Tableau Pulse Features

Now that you know how to enable Tableau Pulse and construct basic metrics, it's time to dig into the more advanced features that are available. These advanced features allow you, the metric author, to create more robust metrics using definition filters, make adjustments to the time dimension, and construct an advanced metric definition with calculated fields or custom filtering. You'll also learn how permissions for metrics work and how to manage them.

Definition Filters

In the preceding chapter, I briefly described a definition filter. A *definition filter* is a preset form of data filtering that limits what records are included in a metric definition. That means for each subsequent metric created from a metric definition, the scope and inclusion of data that is used for the metrics are always filtered in the same way. This behavior is distinctly different from the adjustable metric filters, as they are not visible or accessible (adjustable) to the end user.

A good use case for a definition filter may be when you cannot change or introduce filters in the published data source used for the subsequent metrics. This could be because it is not your responsibility to manage, create, or maintain the published data source. It could also simply mean that introducing those filters for the entire published data source would overly limit and reduce the scope of analysis available. Regardless of the reason, definition filters are the in-between filtering you'll want to use for metric definitions when each downstream metric relies on the same filtered data.

Application of Definition Filters

A definition filter is applied in the same metric definition menu you've seen before. Beneath the measure you select for the metric definition, you'll see a button to add a definition filter. Only text-based (string) fields are available for selection from the dropdown. Let's go through an example with the same published data source used in Chapter 2 (refer to Figure 2-8 for Superstore Transactions).

This time, you'll construct a metric called US Orders, which will be the distinct count of order IDs associated with orders shipped to the US. The data set includes a field called Country/Region that has two values: United States and Canada. You'll be setting the definition filter to United States. Figure 3-1 shows the metric configuration and highlights the location of the Add Definition Filter button.

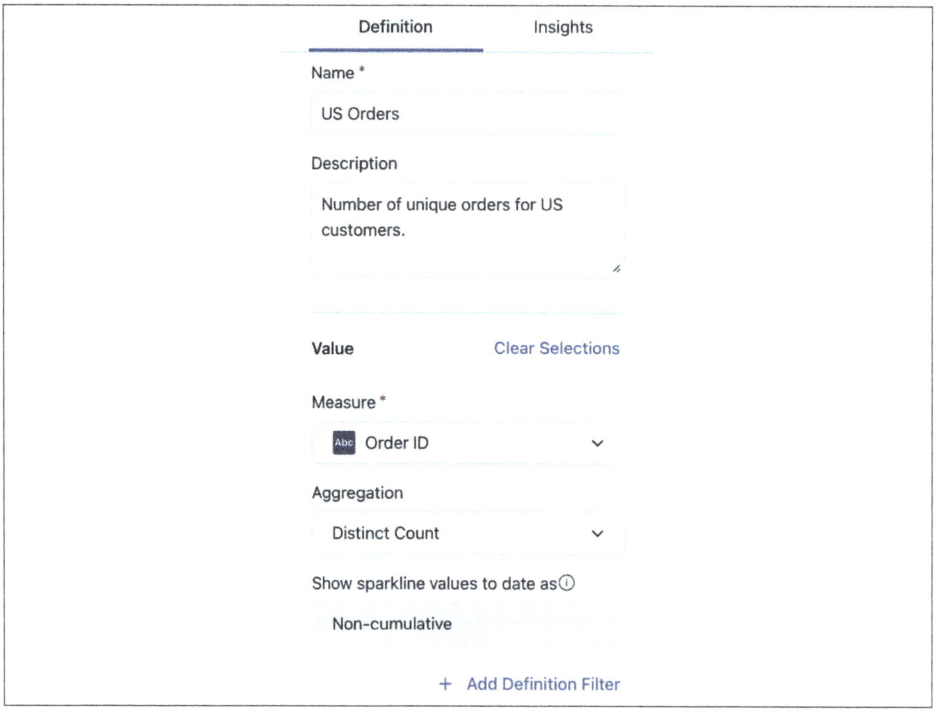

Figure 3-1. The metric definition for US Orders

Clicking Add Definition Filter opens a dropdown where you can select the appropriate field, Country/Region. A window will appear where you can set the behavior of the filter. The filter can be set to include or exclude values in a multiselect format, and there is also a search function for fields with many values. Figure 3-2 shows the options for the field.

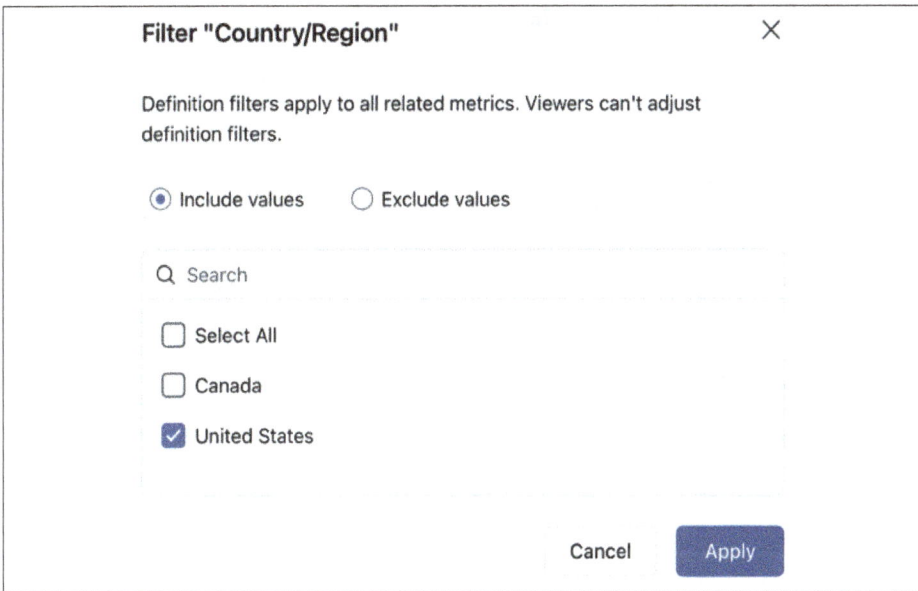

Figure 3-2. *The definition filter for Country/Region*

As you would expect, all the data in the metric definition is now for only the US. From here, if you wanted to, you could apply another definition filter. When you do this, you'll notice that the Country/Region option is no longer available to select. That's because each field's filtering behavior can be configured only once (later in the chapter you'll see how to work around this). Additionally, these filters are immediately applied and considered for any subsequent definition filtering. For example, after selecting United States, applying a filter for State/Province includes only names of US states.

> A best practice for applying filters is to start from the broadest filter, the filter that will reduce the most rows from your data, and continue down to the narrowest filter. Throughout the Tableau platform, filters are typically shown in a vertical list format, so maintaining this approach allows for easy debugging if you end up over-filtering your data or are trying to ensure that only relevant data is included in your analysis.

The great news is that the application of this definition filter applies to all child metrics based off this metric definition. This saves a lot of work and ensures consistency across the metrics.

Relationship to Adjustable Metric Filters

Although definition filters work behind the scenes, that doesn't mean the fields used within those filters can't also be used as adjustable metric filters for end users. Remember, an adjustable metric filter is an interactive filter that end users can access to filter and limit data shown in the metric. The distinction here is that only those values that are included in the definition filter are listed in the subsequent adjustable metric filters. So, if, for example, you include an adjustable filter for Country/Region for the metric definition, as in the previous example, your end user will see only United States as an option for interactive filtering.

The Time Dimension

Chapter 2 covered the time dimension, the field that is used for the date axis and trending. Tableau Pulse also uses the time dimension to compute different comparison periods. The *granularities* (unit of time) that Tableau Pulse supports for dates are days, weeks, months, quarters, and years. It also supports the use of fiscal calendars which have a different start date than the default Gregorian calendar. Remember that fields must exist in the published data source to be selected for the time dimension—as such, modifications can be made to the published data source to explore this functionality.

If you're following along, make these updates to the published data source Superstore Transactions:

1. Create a field called Order Date (Months) by right-clicking the Order Date and creating a custom date field that uses Months as the Detail and the Date Value (see Figure 3-3).

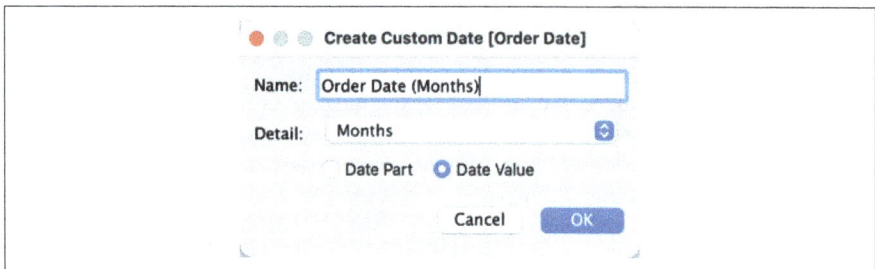

Figure 3-3. The Create Custom Date window

2. Create a field called Order Date (Quarters) by using the same action as the previous step, but this time selecting Quarters as the Detail.

3. Create a field called Order Date (Years), setting Years as the Detail.

4. Right-click Order Date and select Duplicate; rename the field as Order Date FY.

5. Set the Fiscal Year Start for the Order Date FY field to July by right-clicking Order Date FY, and then choosing Default Properties > Fiscal Year Start > July (see Figure 3-4).

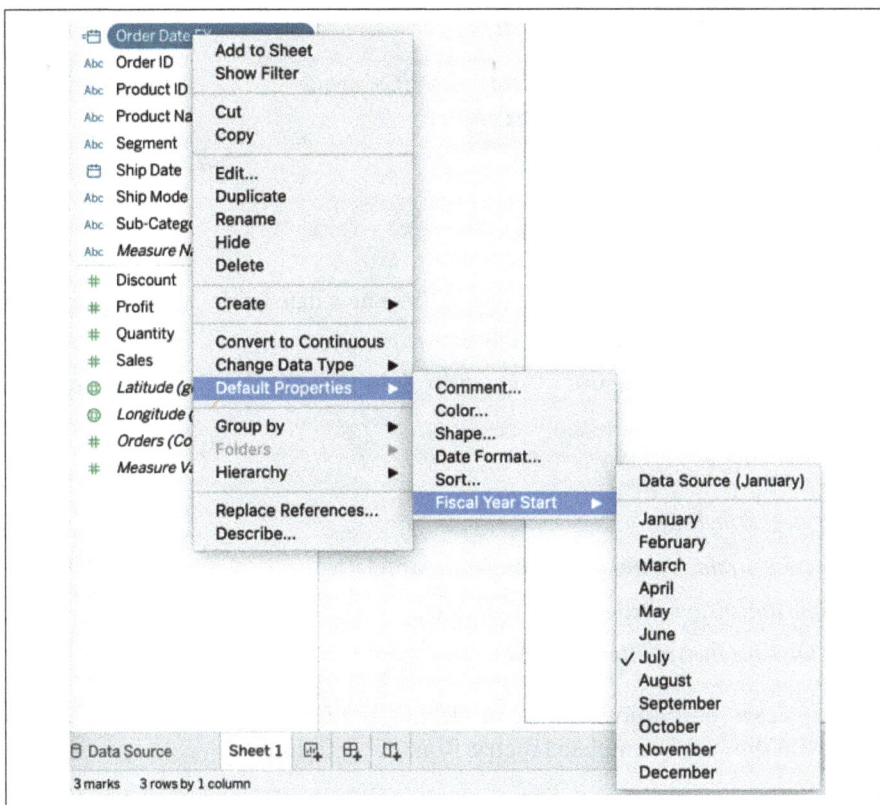

Figure 3-4. Setting the Fiscal Year Start for a field

6. Publish the updated data source to Tableau Cloud.

Now there are four new variations of the Order Date to work with: the first three, which specify the granularity of the date to larger time units (months, quarters, years), and one where the fiscal year starts in July. Let's go through an example with one of these dates to fully explore Pulse functionality.

Exploring Time Granularity

When you created custom dates for months, quarters, and years, behind the scenes Tableau truncated the dates to the specificity you set. Another way to think about it is that it rounded the date to the nearest time unit you specified. Table 3-1 shows how Tableau transforms the date June 15, 2024 (06/15/2024).

Table 3-1. Transformation of date based on specified detail

Date	New date granularity	Transformation
06/15/2024	Months	06/01/2024
06/15/2024	Quarters	04/01/2024
06/15/2024	Years	01/01/2024

You can now create a metric using one of these new date fields; you can use the Quarters version and find out what happens.

This time, construct a metric for Quantity Sold using Order Date (Quarters) by setting the following:

- *Measure*: Quantity

- *Aggregation*: Sum

- *Show sparkline values to date as*: Non-cumulative

- *Time dimension*: Order Date (Quarters)

- *Adjustable metric filters*: Region

You can leave everything else at its default setting and then save the definition. Figure 3-5 shows the completed metric from the definition.

The resulting metric doesn't appear very useful. Instead, it is completely blank, with no metric summary or visualization. Scanning over the metric begins to reveal the *why* behind this. The time dimension chosen is at the quarter level, so the only possible values for the date in use by Pulse are January 1, April 1, July 1, or October 1. Furthermore, the metric identifies that it is Month to Date (next to the Filter button) for the month of June, and the visualization date axis goes from September 1 to September 29. This explains why the metric is blank. There are no September dates to use.

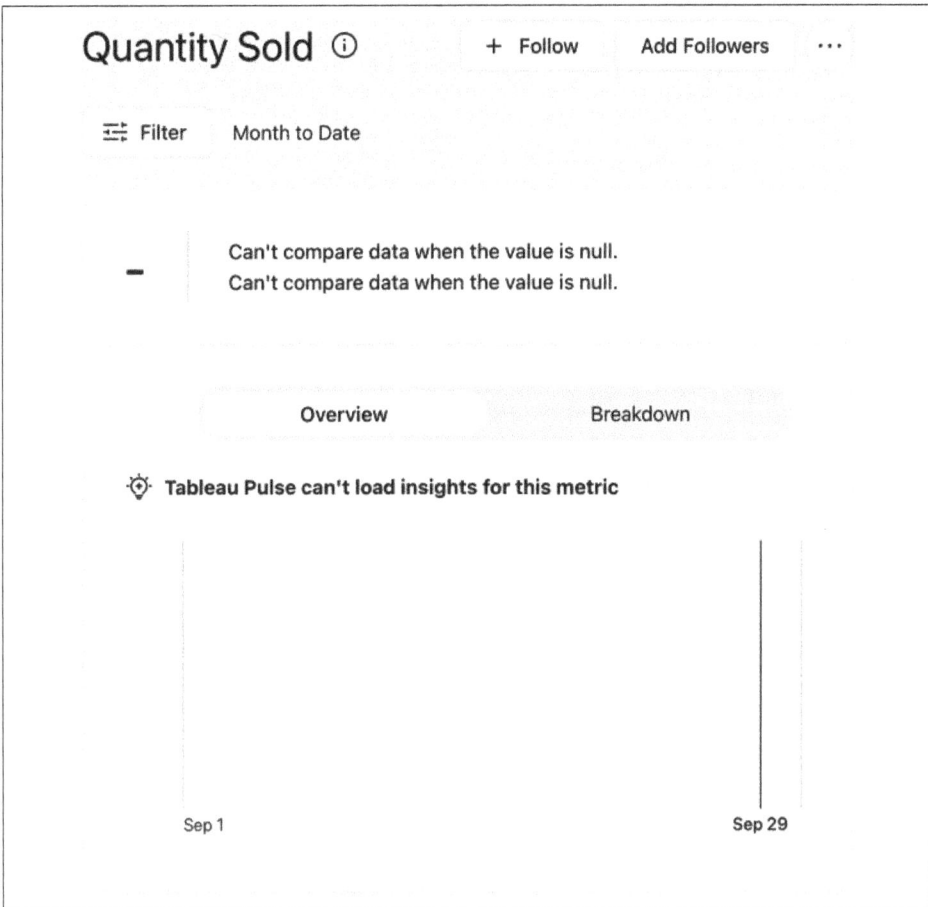

Quantity Sold ⓘ   + Follow   Add Followers   •••

⇶ Filter   Month to Date

—   Can't compare data when the value is null.
           Can't compare data when the value is null.

                    Overview   Breakdown

☼ **Tableau Pulse can't load insights for this metric**

Sep 1   Sep 29

Figure 3-5. Newly created Quantity Sold metric

Setting Minimum Time Granularity

When you encounter this scenario, you can use the Minimum time granularity feature located in the "Advanced time settings" section of the Definition menu. This feature allows you to set the minimum granularity level that the line chart data points will display at; the options are day, week, month, quarter, and year. Once set, any metrics created from the definition will adhere to this behavior, and users will not be able to select time ranges that are *more* granular than the specified granularity. Figure 3-6 shows how applying a default time granularity of quarter impacts the Quantity Sold metric. Notice how the drop-down options for selecting a different time period have been reduced to only those that make sense for dates at a quarter granularity. Similarly, there is a callout when hovering over the dropdown to let the user know this limitation has been set.

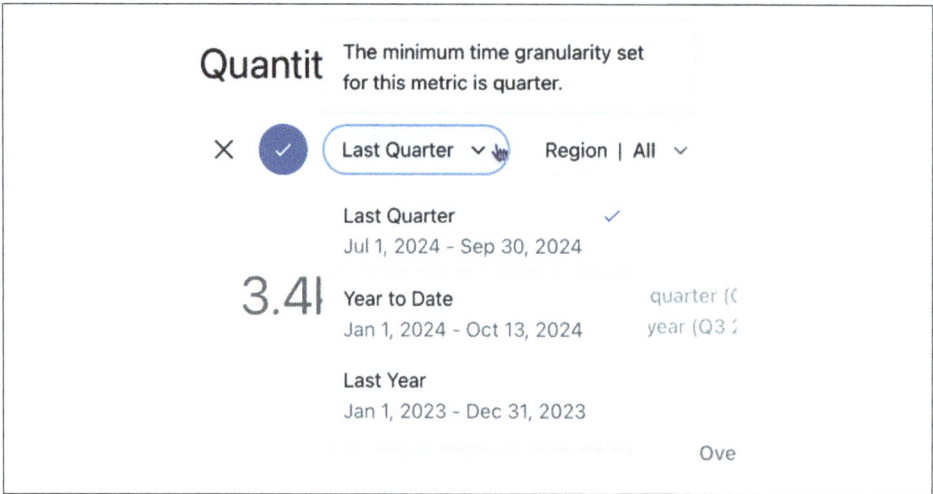

Figure 3-6. Quantity Sold metric time period options after specifying a minimum time granularity

In this metric example, you had a choice to select the less granular date field, but in reality you may not have that choice. As such, it's important to recognize the default time granularity behavior and how specifying a minimum time granularity will impact metrics. The time axis used for each time horizon in Pulse when the minimum granularity is set to the default of days is summarized in Table 3-2 along with how each comparison period is computed. Month to Date is the default time horizon whenever a new metric definition is made.

Table 3-2. Available time horizons with date axis units and comparison periods in Pulse

Time horizon	Date axis units	Prior period	Prior year
Today	Days	The value computed using yesterday	The value computed for the same day from the prior year
Yesterday	Days	The value computed using the day before yesterday	The value computed for the same day (yesterday) from the prior year
Week to Date	Days	The value computed using the prior week	The value computed using the same week from the prior year
Last Week	Weeks	The value computed using the week before the prior week	The value computed using the same week from the prior year
Month to Date (default time horizon)	Days	The value computed using the same number of days of the month directly prior to the current month	The value computed using the same number of days of the same month for the prior year
Last Month	Months	The value computed using the month prior to last month	The value computed using the same month for the prior year

Time horizon	Date axis units	Prior period	Prior year
Quarter to Date	Weeks	The value computed using the same number of days of the quarter directly prior to the current quarter	The value computed using the same number of days of the same quarter for the prior year
Last Quarter	Quarters	The value computed using the quarter before the prior quarter	The value computed using the same quarter from the prior year
Year to Date	Months	The value computed using the same number of days for the prior year	No comparison (identical to prior period)
Last Year	Years	The value computed for the year prior to last year	No comparison (identical to prior period)

From the table, you can see that many of these time horizons display metrics using days, so it's safe to say that you'll get the most benefit out of Pulse if your time dimension is at a daily granularity.

> Tableau Pulse has recently introduced the ability to specify the lowest date granularity allowed for a metric. However, your author still believes you'll get the most flexibility with sticking to days.

Using Fiscal Calendars

Tableau Pulse also supports the use of fiscal calendar dates for the time dimension. As mentioned earlier, a *fiscal calendar* has a different start date than the Gregorian calendar (which starts in January). You already set the fiscal start of the duplicated Order Date field, Order Date FY, to July, so now it's time to see it in action.

Construct a metric called Profit from Sales as follows:

- *Measure*: Profit
- *Aggregation*: Sum
- *Show sparkline values to date as*: Running total
- *Time dimension*: Order Date FY
- *Adjustable metric filters*: Category

You'll notice that, beneath the time dimension box, a new message states that the fiscal year start for the field is July, as seen in Figure 3-7.

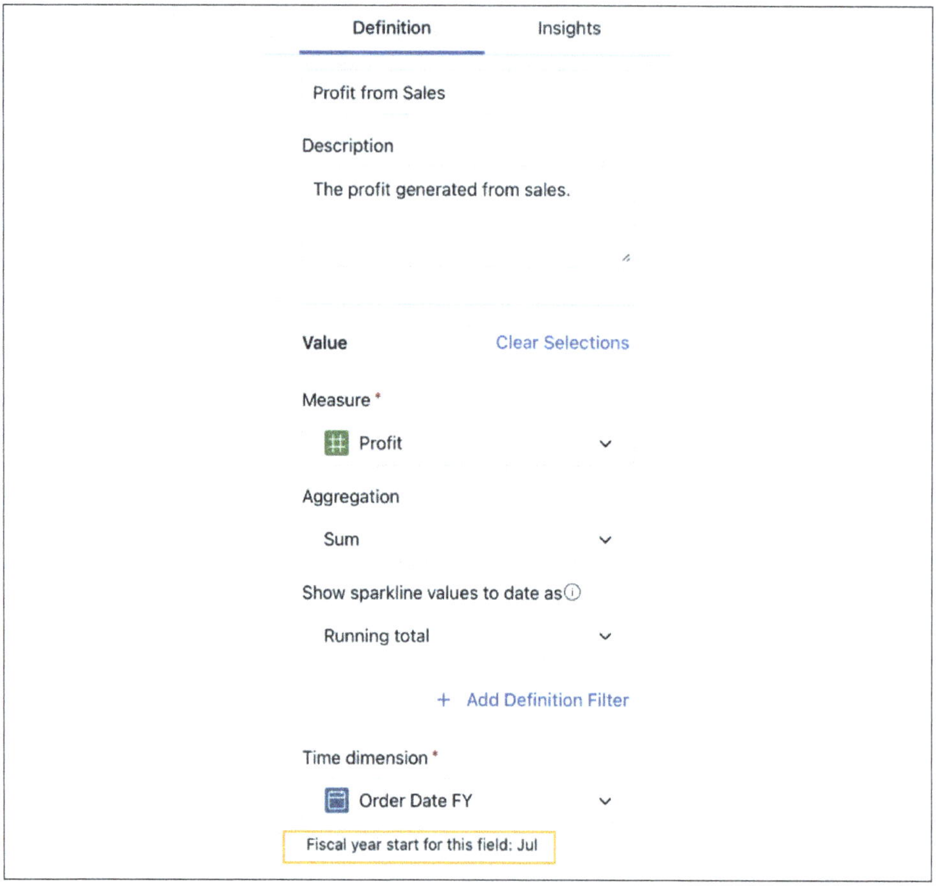

Definition   Insights

Profit from Sales

Description

The profit generated from sales.

Value   Clear Selections

Measure *

▣ Profit   ˅

Aggregation

Sum   ˅

Show sparkline values to date as ⓘ

Running total   ˅

\+   Add Definition Filter

Time dimension *

📅 Order Date FY   ˅

Fiscal year start for this field: Jul

Figure 3-7. The fiscal year start highlighted for the selected time dimension

Tableau Pulse will now use this information to construct the time horizons and comparison periods. It also includes the word *fiscal* in the summaries and time horizon options. The scope and size of the time horizons don't change with the use of a time dimension, and neither do the date axis and time units. Only the names of the periods, such as Fiscal Quarter to Date instead of Quarter to Date, are modified. Figure 3-8 shows the metric when set to Last Fiscal Quarter.

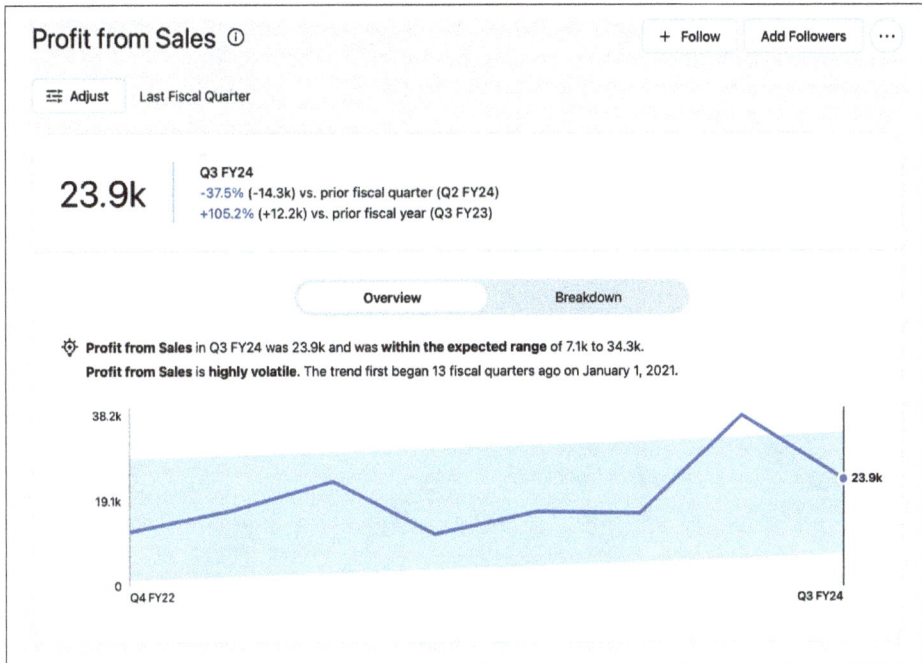

Figure 3-8. A metric using a time dimension with a fiscal calendar start of July (see a larger version of this figure online (https://oreil.ly/lait0308))

Offsetting the Date

You can adjust one last setting related to the time dimension, which is an offset of days from today. This can be very useful if the underlying data for the metric you're constructing isn't available through today. It can also be useful if the data for today is in flux and including it in the metric could generate misleading insights. This setting is accessed by clicking "Advanced time settings" beneath the "Time dimension" dropdown in the metric Definition. Clicking this opens an input box where you can enter a positive integer from 0 to 365 days for the offset. This offset works for both fiscal and default time dimensions. Figure 3-9 shows the input box for the metric from the previous example. When an offset is specified, the metric will no longer include data from the days included in the offset. Functionally, this means when you select Month to Date, Quarter to Date, or Year to Date time horizons, the end date for each of these selections will be today minus the offset days.

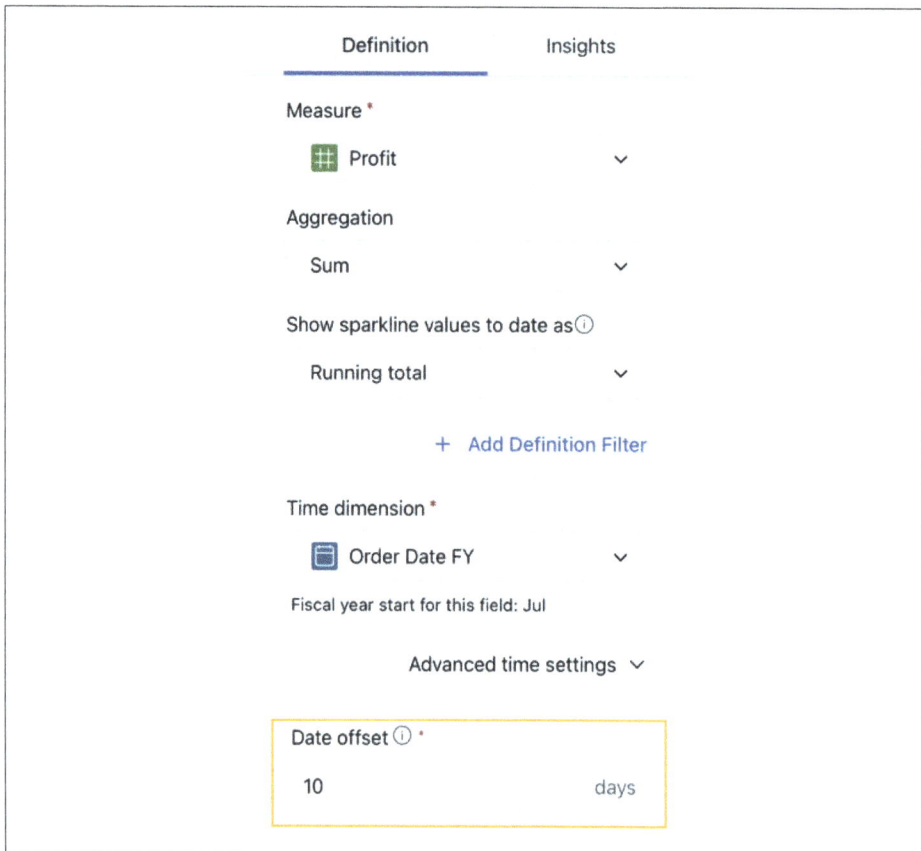

Figure 3-9. The Date offset box for a time dimension

Advanced Metric Definition

Up to this point you've relied on preexisting fields in the published data source to construct Pulse metrics. But what happens if you need to create a metric that relies on a calculation based on two measures in your data? What if you need a definition filter more robust than a multiselect list, or you simply prefer to work with the worksheet editor interface you're accustomed to in the Tableau visualization authoring experience? In these scenarios, using the advanced metric definition feature will be necessary. The *advanced metric definition* functionality grants you access to an editor interface where you can drag and drop fields onto a canvas. Additionally, you'll have access to the majority of the functionality (like creating bins, sets, and calculated fields) that you're used to in the typical authoring experience. When you choose to use this feature, it will replace the fields in the metric definition.

> When using the advanced metric definition functionality, you'll be specifying all three components of the metric definition: the measure, the time dimension, and definition filters (optional).

Calculated Field as Measure

It's easy to imagine that you may need to create a new field based on other fields in your existing data set. Often transformation calculations like ratios, normalizations, or logic statements can add deeper and more impactful analysis and insight to your data. These fields are unlikely to be precomputed in your data source (especially if you're connecting directly to the source system). Here you have a choice in approach: construct the calculated field in the published data source and then select it in the metric definition as the measure, or calculate the field by using the advanced metric definition feature. Let's walk through the latter.

Using the same published data source, Superstore Transactions, you'll construct a new field called Profit Ratio. Figure 3-10 shows the Advanced Analytics Editor that pops up when you select Advanced Metric Definition located below the "Compared to" section in the Definition menu.

Here you see the familiar worksheet editor with a few minor modifications. The Data pane is still available on the left for access to the fields in your data source. However, instead of having a Rows shelf and Columns shelf, you have a Measure shelf and Time Dimension shelf. There is still a Filters shelf, but the Marks card is completely missing. Additionally, in this editor you are not able to control the type of visualization that is shown in the canvas.

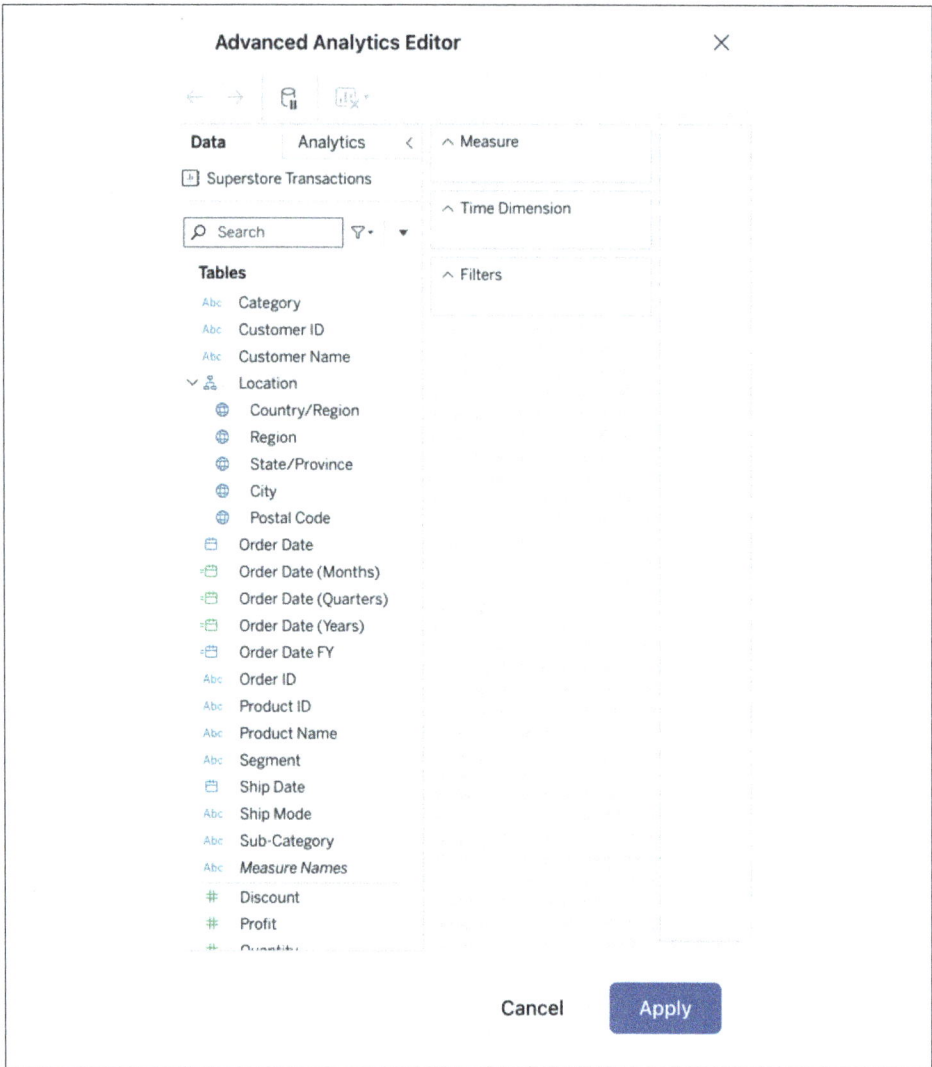

Figure 3-10. The Advanced Analytics Editor

To construct the Profit Ratio calculation, right-click the Profit metric, and choose Create > Calculated Field as shown in Figure 3-11.

This action displays a calculation editor where you can enter in the following calculation: **SUM(Profit) / SUM(Sales)**; name the new field Profit Ratio, as shown in Figure 3-12.

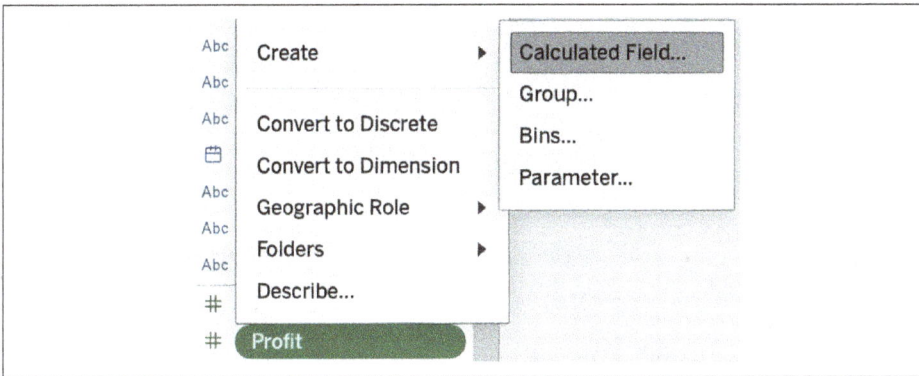

Figure 3-11. Creating a calculated field from the Profit field

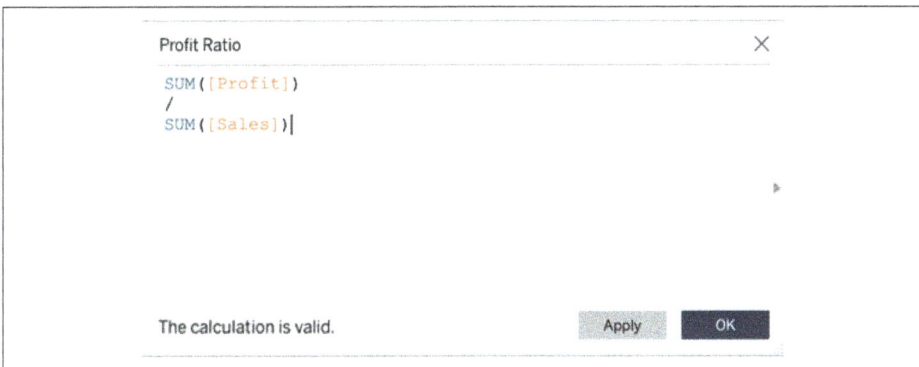

Figure 3-12. A calculation for Profit Ratio

This new field can now be dragged onto the Measure shelf. Next, drag Order Date onto the Time Dimension shelf. By default, the Order Date will show up as a *discrete* dimension with the YEAR() function—carryover default behavior from the typical visualization authoring experience. A *discrete* field in Tableau has a finite list of options for the field value and is visually denoted by the blue color of the field. Additionally, discrete fields are used as headers in visualizations. This is in contrast to a *continuous* field that shows up as green and generates a visualization axis, demonstrating the infinite possibilities of potential values. The YEAR() function returns the four-digit year associated with the date. The resulting visualization is shown in Figure 3-13. Paradoxically, the behavior applied to the Order Date field in the visualization is completely ignored when passed back to the metric definition. This means that although the visualization in the Advanced Analytics Editor shows a chart of Profit Ratio by the year of the Order Date, the metric definition and subsequent metrics will access the Order Date by its natural time granularity of days.

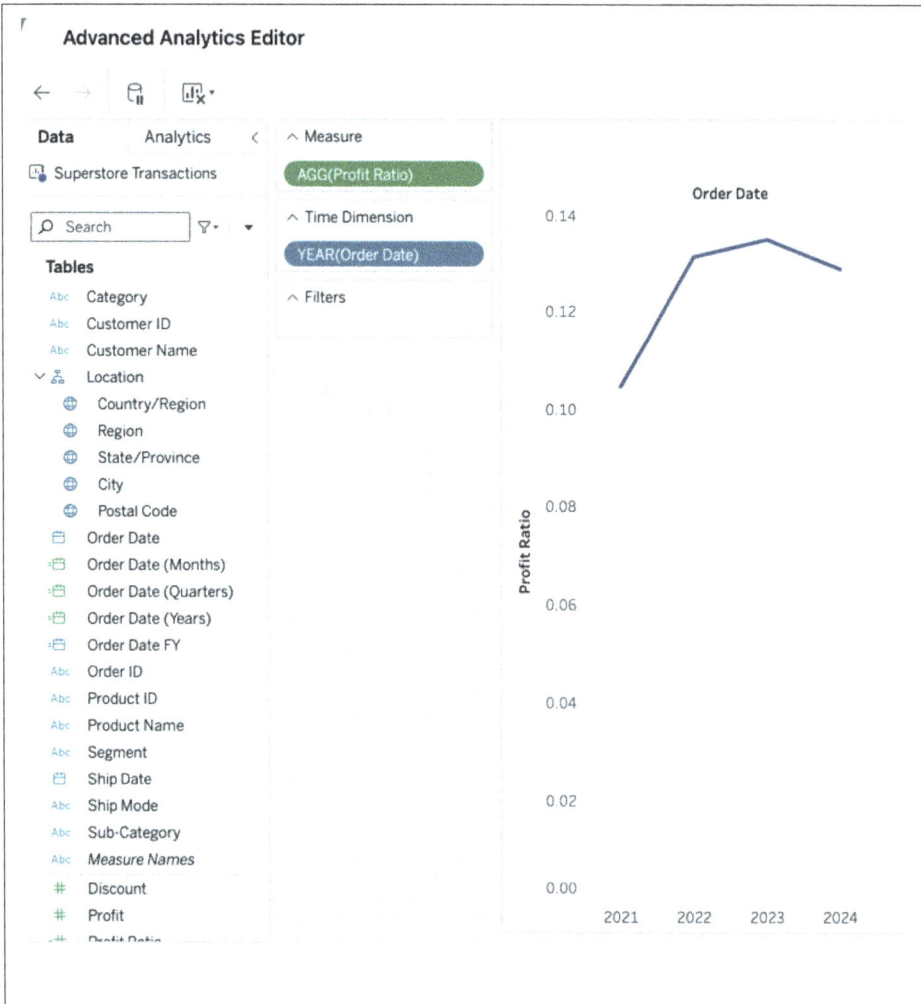

Figure 3-13. The Advanced Analytics Editor with the Profit Ratio calculated field in use

When you click Apply to save the advanced metric definition, a warning message pops up reminding you that there must be a measure and a time dimension specified for the definition to be valid, as shown in Figure 3-14. If your advanced definition is missing any of the required components, you'll be unable to exit the editor screen, and a red error message will display at the top.

Figure 3-15 shows how the selections from the editor are mapped to the sections in the metric definition. When a calculated field is chosen for a Measure, the option to specify a running total sparkline for *to date* options is no longer available and instead will always be displayed as non-cumulative.

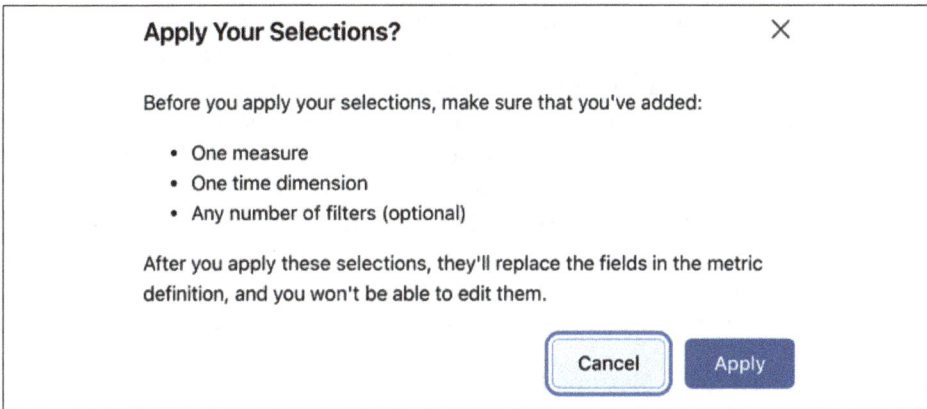

Figure 3-14. *Warning message when applying an advanced metric definition*

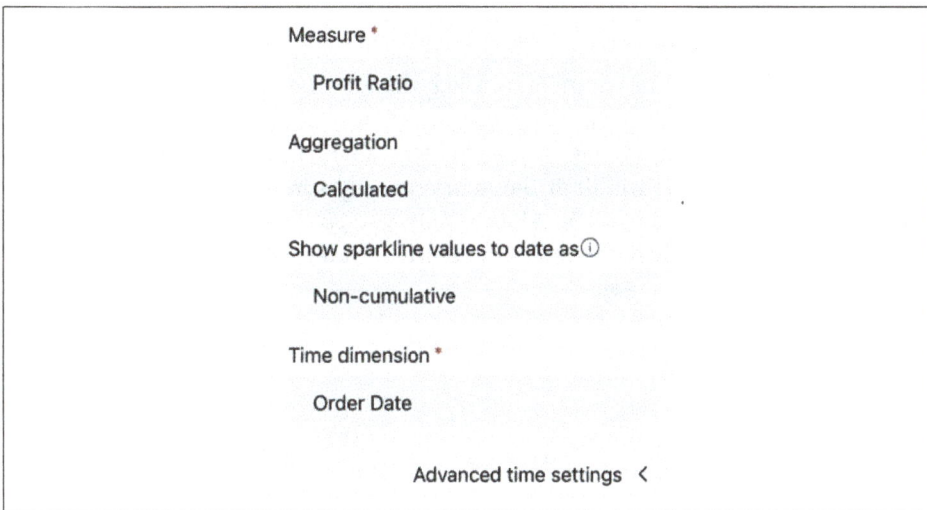

Figure 3-15. *The metric definition generated from the advanced metric definition constructed in the Advanced Analytics Editor*

Custom Definition Filter

One of my favorite features of Tableau is the ability to filter data in a variety of ways. As you saw earlier in this chapter, text fields can be used as definition filters by way of a multiselect include or exclude list. However, discrete dimensions of any field type can also be filtered using a *wildcard* condition, a condition that relies on the value of an aggregated field or formula, and a *top* or *bottom* condition that relies on the value of an aggregated field or formula. All these additional filtering types can be applied only as definition filters through the Advanced Analytics Editor. The editor is also the only place you'll be able to apply a definition filter based on a measure.

Imagine you get a request to construct a metric that displays the quantity of Canon printers, copiers, scanners, and fax machines sold. You may consider building a multiselect definition filter based on the Product Name field in the traditional way, but this has limitations. First, there are several values, and it can be easy to miss one in a large checklist. Next, new products are added to inventory frequently. If you use the typical multiselect list, new values won't automatically be added to the filter. Instead you'll end up maintaining the definition filter in perpetuity, as you try to ensure all past and future products are included in the filter. It would be better to construct a wildcard filter based on Product Name. To do this, click the Create Advanced Definition button in a new metric definition window. Then do the following:

- Drag Quantity to the Measure shelf.
- Drag Order Date to the Time Dimension shelf.
- Drag Product Name to the Filters shelf.

Once you drag the Product Name to the Filters shelf, the full filter configuration window will appear, as seen in Figure 3-16. Click the arrow next to General to collapse the multiselect list functionality, then click the arrow next to Wildcard to expand the options. From here, enter "**Canon**" in the "Match value" box, and keep the Comparison set to Contains. This will limit the metric to only sum the quantity of products sold that contain the name *Canon*. Alternatively, you could select "Starts with" if you wanted to enforce that the first word of the Product Name is *Canon*.

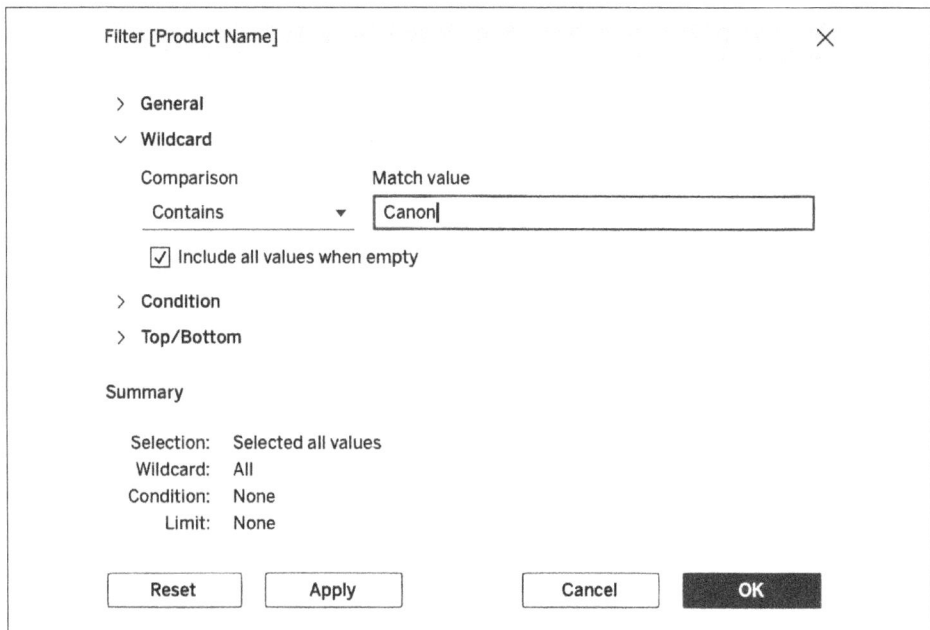

Figure 3-16. Constructing a custom wildcard definition filter

To enhance this metric further, you can add Product Name as an adjustable filter in the definition. When metrics from this definition are constructed, end users can select from the dropdown of available Canon products and see them in the Breakdown. Figure 3-17 shows the newly created metric and breakdown for the Year to Date.

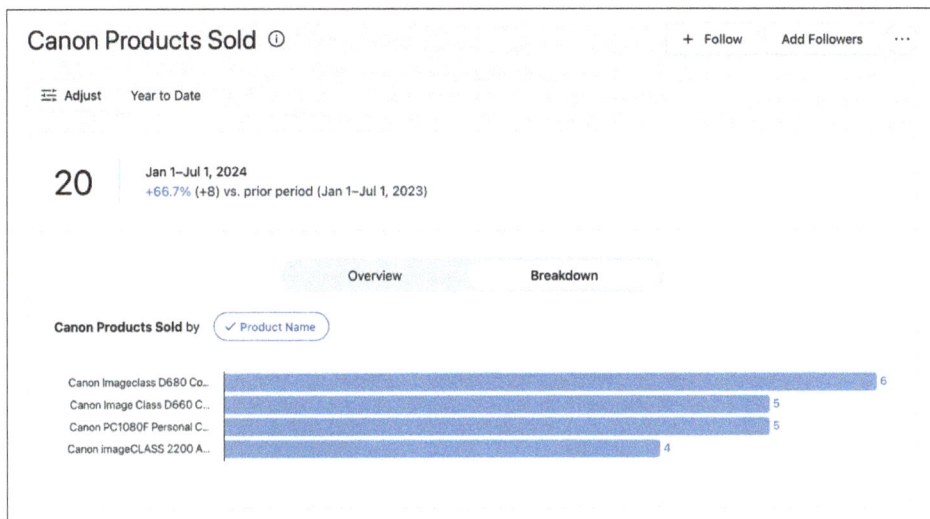

Figure 3-17. Metric with Breakdown shown after applying custom definition filter (see a larger version of this figure online (https://oreil.ly/lait0317))

Now let's go through an example where the metric has a definition filter based on a measure. This type of filtering is useful anytime you want to isolate data used in a Pulse metric based on a numerical value.

Continuing with Superstore Transactions, you can construct a metric showing the number of items sold where the Discount value was at least 50%. At Superstore, any sale with a discount of 50% or greater represents discontinued products. The department buyers like to track these transactions to minimize the amount of unpopular products stocked. To do this, click the Create Advanced Definition button in a new metric definition window. Then do the following:

- Drag Quantity to the Measure shelf.
- Drag Order Date to the Time Dimension shelf.
- Drag Discount to the Filters shelf.

Once you drag Discount to the Filters shelf, a menu will pop up (Figure 3-18). Since this is a numerical filter, you're given the option to filter by the individual values that exist in the Discount field ("All values"), or you can filter by a specified aggregation. Additionally, because you want to filter individual rows of data that have a discount value of at least 50%, select "All values." An additional box will display (Figure 3-19) that allows you to specify the way the data is limited. Here choosing the "At least" option and setting the minimum to 0.5 (50%) makes the most sense.

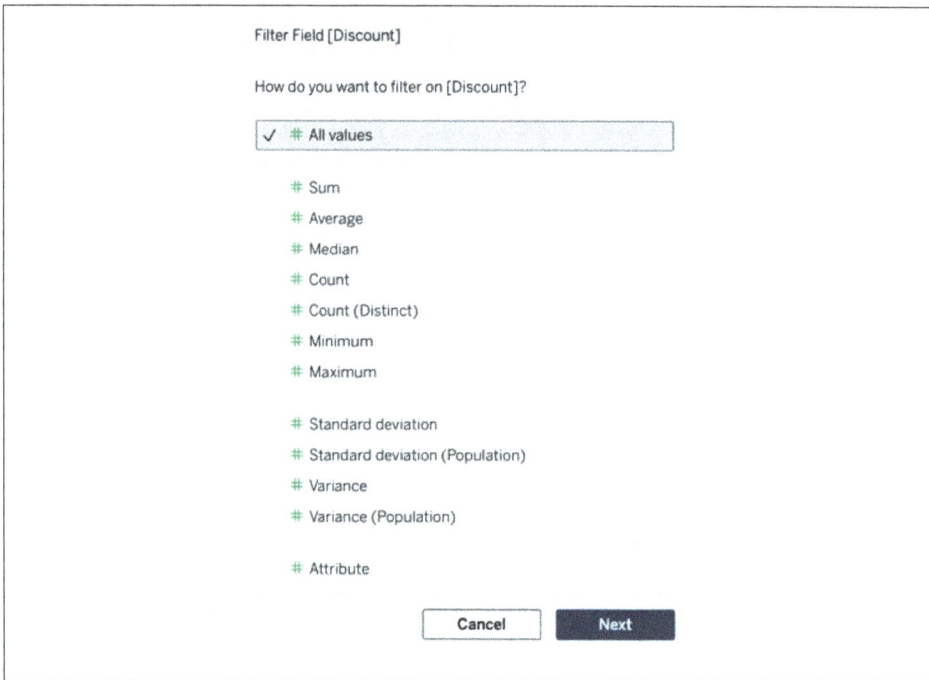

Figure 3-18. Filter menu specifying how a measure is to be filtered

As a last step, include Sub-Category as an adjustable filter. Remember, the department buyers are on the lookout for products that aren't sold until they are discontinued, so including the Sub-Category will help them identify if one particular type of product isn't very popular. Figure 3-20 shows the completed metric set to Quarter to Date with Breakdown; it appears that many Binders have been discontinued and sold.

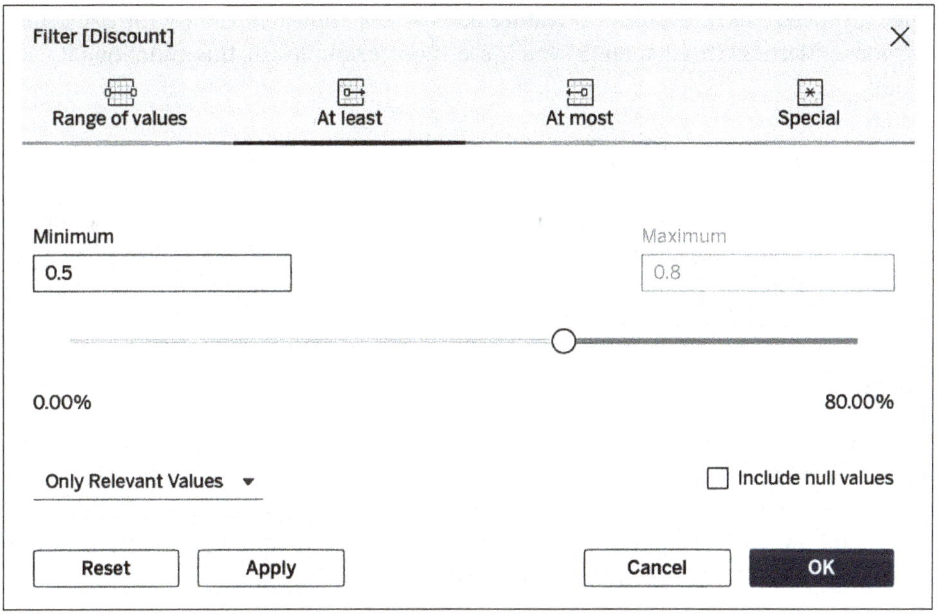

Figure 3-19. Filter menu specifying the minimum Discount value to be included

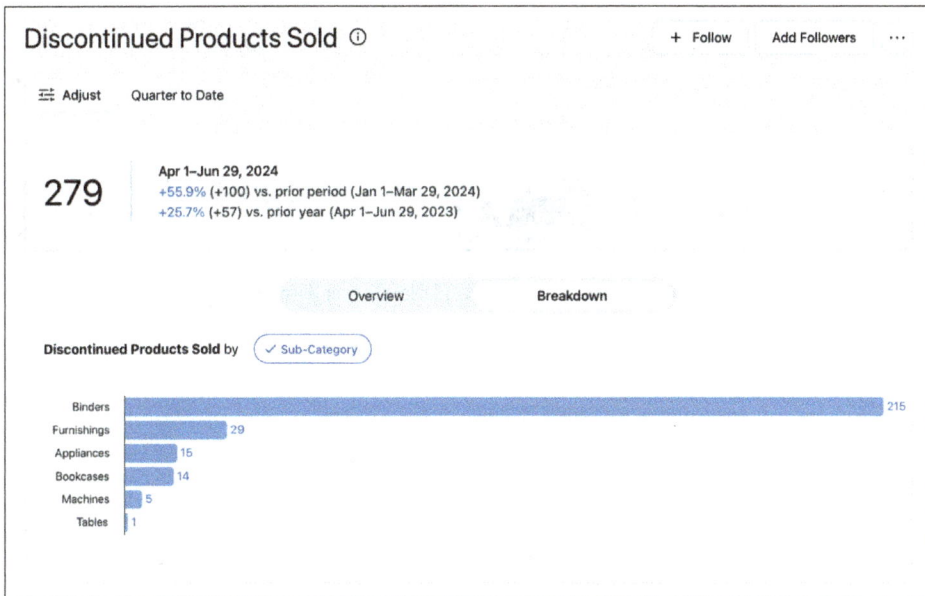

Figure 3-20. Metric with a custom definition filter based on a measure (see a larger version of this figure online (https://oreil.ly/lait0320))

The Advanced Metric Definition feature adds a very robust capability for constructing Pulse Metrics. In Chapter 5, you'll see more examples of this functionality in action.

Managing Pulse Metrics

Now that you've seen how to construct advanced Pulse metrics, it's time to turn your attention to managing the metric definitions and subsequent metrics that are created. Remember that one of the most important responsibilities you take on is ensuring that the right people have access to the right information.

Permissions

The Tableau platform has a very detailed and robust permissions capability. Permissions determine not only who has access to what, but also the level of interactivity an end user can have, including whether they can see disaggregated data. In addition to permissions, content is organized in folders called *Projects* where permissions can be applied en masse to all content contained within the folder.

Permissions for Pulse metrics start at the published data source level. This alone will determine whether a user will be able to create or follow a metric. For each published data source, several permission types are available (described in Table 3-3). It's important to remember that these permissions are set in Tableau Cloud and won't be accessible or adjustable when in Pulse.

Table 3-3. Types of permissions available for a published data source

Permission	Granted ability	Necessary for Pulse
View	User can see the data source on Tableau Cloud.	Yes
Connect	User can connect to the data source by using Tableau Desktop, Prep Builder, Ask Data, or web authoring.	Yes
Download Data Source	User can download published data source as a standalone file (*.tdsx*).	No
Overwrite	User can overwrite the existing data source.	No
Save As	User can save the published data source as a new file.	No
Move	User can move the data source to a different project.	No
Delete	User can delete the data source.	No
Set Permissions	User can create and edit permissions rules for the data source.	No

To create or access Pulse metrics, a user must have both the View and Connect permissions. In addition to permissions, capabilities of a user are also governed by their site role. There are six roles:

- Site Administrator Creator

- Creator

- Site Administrator Explorer

- Explorer (can publish)

- Explorer

- Viewer

Viewers and Explorers will be able to create and access metrics, whereas Site Administrator Creators, Creators, Site Administrator Explorers, and Explorers (can publish) will be able to create metric definitions along with metrics. Finally, permissions can be applied to user groups or individual users. Figure 3-21 shows the permissions assigned to a group for the Superstore Transactions published data source. These permissions will allow anyone in the group to access and build metrics from definitions, and those with an Explorer (can publish) or greater role can create new metric definitions.

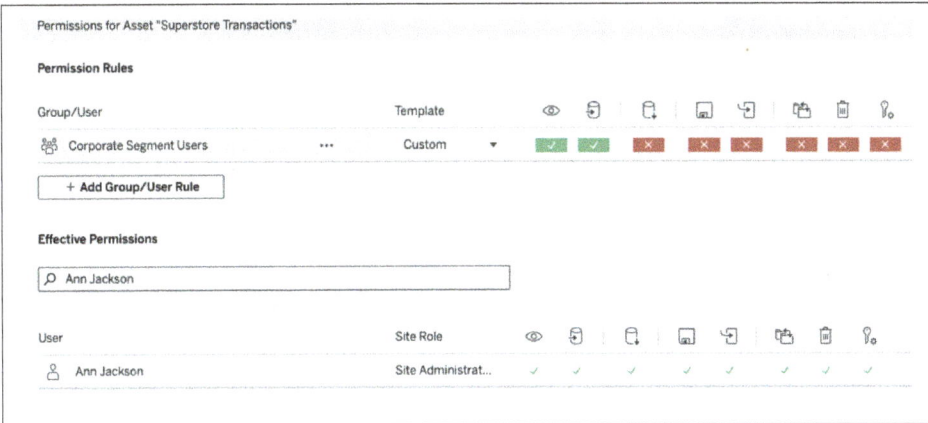

Figure 3-21. Permissions set for published data source (see a larger version of this figure online (https://oreil.ly/lait0321))

Followers

To see what metric definitions and metrics exist in Pulse, go to its Browse Metrics section. Remember, you'll need to have a Site Administrator Creator, Creator, Site Administrator Explorer, or Explorer (can publish) role to edit or delete the metric definition. Figure 3-22 shows the metric definitions on my Tableau Cloud site.

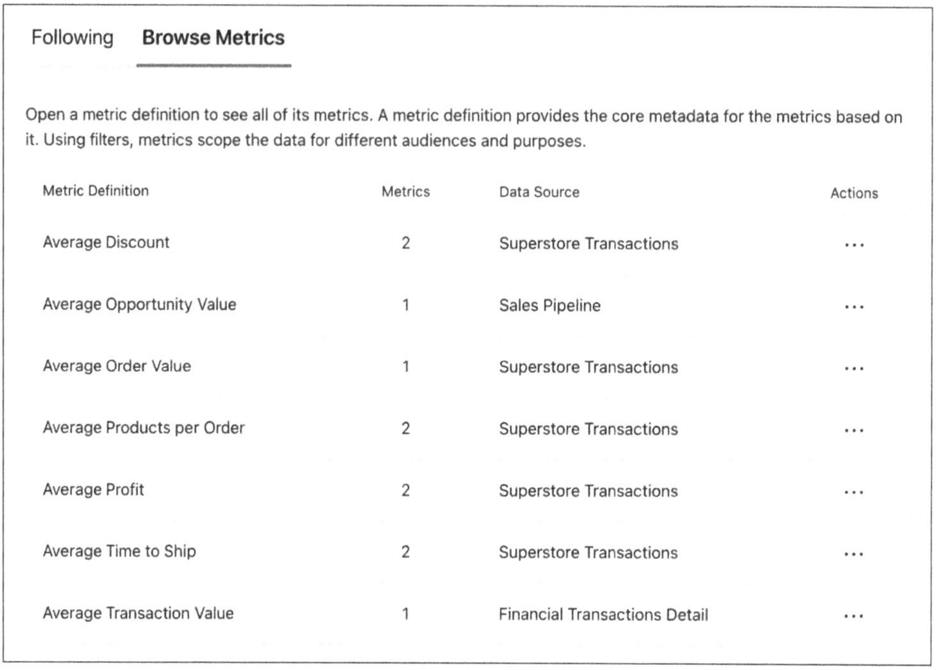

Following   **Browse Metrics**

Open a metric definition to see all of its metrics. A metric definition provides the core metadata for the metrics based on it. Using filters, metrics scope the data for different audiences and purposes.

Metric Definition	Metrics	Data Source	Actions
Average Discount	2	Superstore Transactions	...
Average Opportunity Value	1	Sales Pipeline	...
Average Order Value	1	Superstore Transactions	...
Average Products per Order	2	Superstore Transactions	...
Average Profit	2	Superstore Transactions	...
Average Time to Ship	2	Superstore Transactions	...
Average Transaction Value	1	Financial Transactions Detail	...

Figure 3-22. List of metric definitions and metrics available

Within the screen in Figure 3-22, you can see the name of metric definitions, the number of child metrics created for each definition, and the data source the definition is connected to. An Actions menu allows you to navigate to the Insights Exploration section, see all metrics created from the definition, and both edit and delete the definition. Figure 3-23 shows the metrics that exist for the Sales metric definition.

It is from this screen that you'll be able to see and manage who is following specific metrics. Each metric constructed indicates the number of followers and whether you are following the metric, and has a small triple dot actions menu to manage the followers. From this menu you can see, add, and remove groups and users from a metric. Figure 3-24 shows the followers for the Quarter to Date - Corporate metric.

Depending on where you're looking, the number of followers may show as two different values. Within the Manage Followers menu, it is the count of unique groups and individual users named who follow the metric. On the metric itself (in Figure 3-23), the number of followers represents the unique number of users and is therefore the accurate value to use to track the number of people following a given metric.

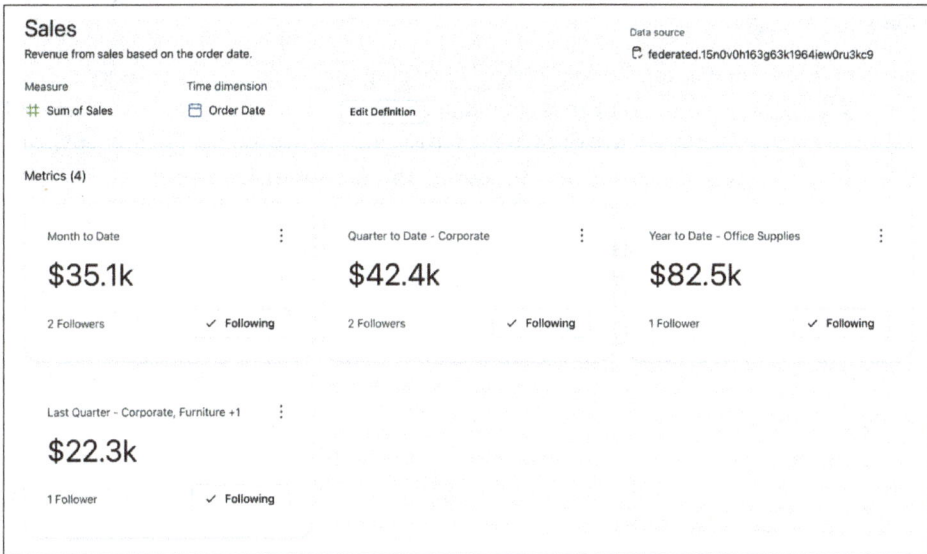

Figure 3-23. Metrics associated with the Sales metric definition (see a larger version of this figure online (https://oreil.ly/lait0323))

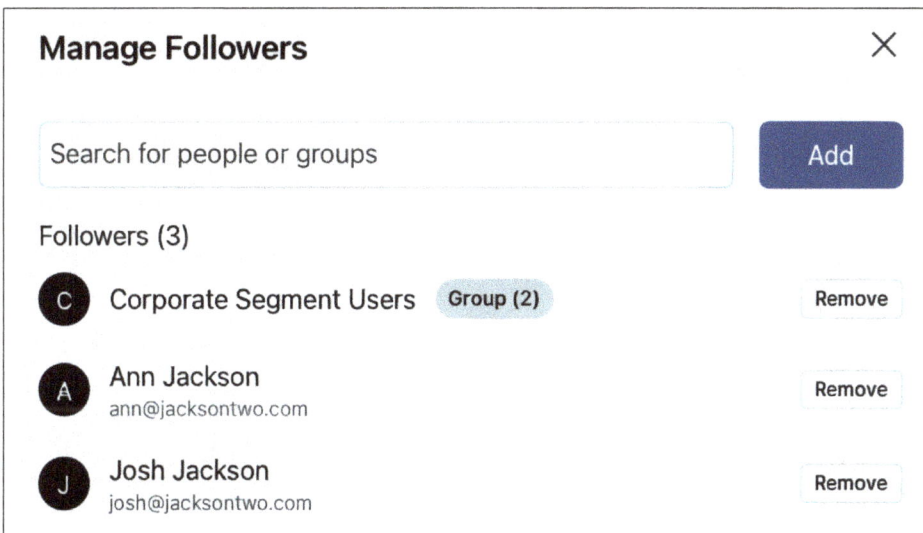

Figure 3-24. The followers assigned to a metric

If a user's access to a metric is in question, first check their permissions for the published data source by going to the permissions of the data source and entering the user's name in the Effect Permissions section. This will also show you their site role. Figure 3-21 shows my effective permissions.

Tracking Usage

The last piece of managing Pulse metrics is tracking the quantity of definitions, metrics, and users. This is different from managing *who* is following a metric and instead aims at seeing a holistic picture of metrics across your entire Tableau Cloud environment. To do this, you can access data sources in the Admin Insights project.

> Admin Insights is a project automatically created for each Tableau Cloud site. It includes a prebuilt workbook and data sources that contain metadata usage and information about content on Tableau Cloud and Pulse.

You can use two data sources within Admin Insights to manage Pulse metrics. The first is the Site Content data source. This data source includes the content's type, name, and key features. The most useful fields from this data source for managing Pulse metrics are the following:

Created At
> The UTC timestamp (date and time) when an item was created

Created At (Local)
> The timestamp (date and time), controlled by the Timezone parameter, when an item was created in your local timezone

Item Type
> The type of content for Pulse items, either metric definition or metric

Item Name
> The name of the item

Owner Email
> The email address of the user who created the metric

Description
> The description listed in the metric definition

Adjustable Filter Options
> The fields added and selected as adjustable filter options for the metric definition

Has Definition Filters
> A Boolean denoting whether definition filters are applied to a metric definition

Metric Definition Aggregation
> The aggregation of the measure tracked in the metric definition

Metric Definition Data Source ID
> The data source that the metric definition is connected to

Metric Definition ID for Related Metric
    The definition that the metric originates from

Metric Definition Number Format
    The number format of the metric definition value

Metric Time Granularity
    The level of detail that the metric is aggregated to

The other useful data source is called Permissions. This data source lists the permissions type and granted permissions for all content on Tableau Cloud. Additionally, it includes whether the grantee is a user or a group. The most useful fields from this data source for managing Pulse metrics are the following:

Capability Type
    The type of action a user can perform on a piece of content, like view or connect

Permission Description
    The detailed description of the capability

Permission Value
    A numerical representation of the effective permission

Grantee Name
    The name of the user or group with permissions specified

Grantee Type
    Denotes whether the grantee is a user or a group

Item Name
    The name of the piece of content the permissions are applied to

Item Type
    The type of content

User Email
    The email address of the user with permissions

Has Permission?
    A calculated Boolean field that determines whether a user has the stated permission.

In Chapter 5 you'll see examples of Pulse metrics that can be created from these data sources for easy administration.

Summary

In this chapter, you've had exposure to the advanced features and capabilities in Tableau Pulse. Here are some key takeaways you've learned:

- Definition filters limit the scope of data included in a metric definition and its subsequent child metrics.
- The time dimension can have a granularity of day, week, month, quarter, or year.
- For the most benefit, use a time dimension with granularity of a day.
- Pulse will use fiscal calendar information assigned to a time dimension to adjust the time horizons and language present in the metric.
- You can offset the end date of a pulse metric by using the advanced time settings.
- The advanced metric definition feature allows you to create calculated fields and use more robust filtering options.
- Using the advanced metric definition feature replaces the measure, aggregation, time dimension, and definition filters shown in the Value section of the metric definition.
- Custom definition filters can be made using the advanced metric definition that include more robust filtering such as wildcard filtering or numerical measure filtering.
- Permissions for metrics are governed by the permissions a user has for the published data source associated with the metric.
- The followers for a metric can easily be managed through the action menu associated with the metric.
- Two published data sources in Admin Insights contain metadata information for tracking usage and permissions associated with Pulse.

In the next chapter, you'll take a step back and learn what the Pulse environment looks like for an end user (not a metric author). You'll see how their Pulse summaries are organized and displayed across a variety of environments, including Tableau Mobile. Finally, you'll see how they can set up digests via email or Slack and what those digests consist of.

CHAPTER 4

Tableau Pulse for End Users

In previous chapters, you learned how to enable Tableau Pulse and build both basic and advanced metric definitions. Now it's time to see how they're consumed from your audience's perspective. In this chapter, you'll see how metrics you create are displayed to end users, what happens when they follow multiple metrics, and how they can receive summarized digests of the AI insights generated.

Pulse Home Page

Just as you saw in Chapter 2, after navigating to Tableau Pulse from Tableau Cloud, end users are greeted with a summary page that shows the metrics they already follow and those that are available for following. However, instead of being able to see all metrics in Pulse, remember they will have access to only those tied to published data sources that they have View and Connect permissions for. Figure 4-1 shows the available metric definitions for the Viewer role with access to the published data source used in previous examples, Superstore Transactions.

From this screen, users will access all metric definitions and existing metrics and begin the process of constructing their own metrics. By clicking the triple-dot Actions menu on the right, they'll be able to access two subsequent areas:

Insights Exploration
 The summary and breakdown screen for the metric definition

See All Metrics
 A screen with all existing metrics based on the metric definition

They can also click the metric definition to navigate to the screen with all metrics.

Open a metric definition to see all of its metrics. A metric definition provides the core metadata for the metrics based on it. Using filters, metrics scope the data for different audiences and purposes.

Metric Definition	Metrics	Data Source	Actions
Average Discount	2	Superstore Transactions	...
Average Order Value	1	Superstore Transactions	...
Average Products per Order	2	Superstore Transactions	...
Average Profit	2	Superstore Transactions	...
Average Time to Ship	2	Superstore Transactions	...
Canon Products Sold	1	Superstore Transactions	...
Discontinued Products Sold	1	Superstore Transactions	...
East Region Sales	1	Superstore Transactions	...
Items Sold	2	Superstore Transactions	...
Monthly Sales	1	Superstore Transactions	...

Figure 4-1. Available metric definitions for a Viewer role

Once again, this should all seem familiar, as the actions available are a subset of those you saw in Chapter 3. However, for a Viewer, they can view (hence the name) but cannot modify or delete any metric definitions they have access to.

> Although not shown in the figures, if your Tableau Cloud site has the Data Catalog enabled, data quality warnings applied to data assets will show up in the Browse Metrics section. They will also show up on the bottom of a followed metric. Finally, they appear beneath the metric's name when viewing either the Overview or Breakdown.

Figure 4-2 shows the available metrics based on the Sales metric definition that was constructed in Chapter 2. You'll notice that the Viewer cannot see the number of followers for each metric, and as expected, they aren't able to edit the metric definition.

While Viewers of metrics aren't able to delete or modify them, they are able to manage the followers. They can both view the existing followers as well as add and remove them from a given metric.

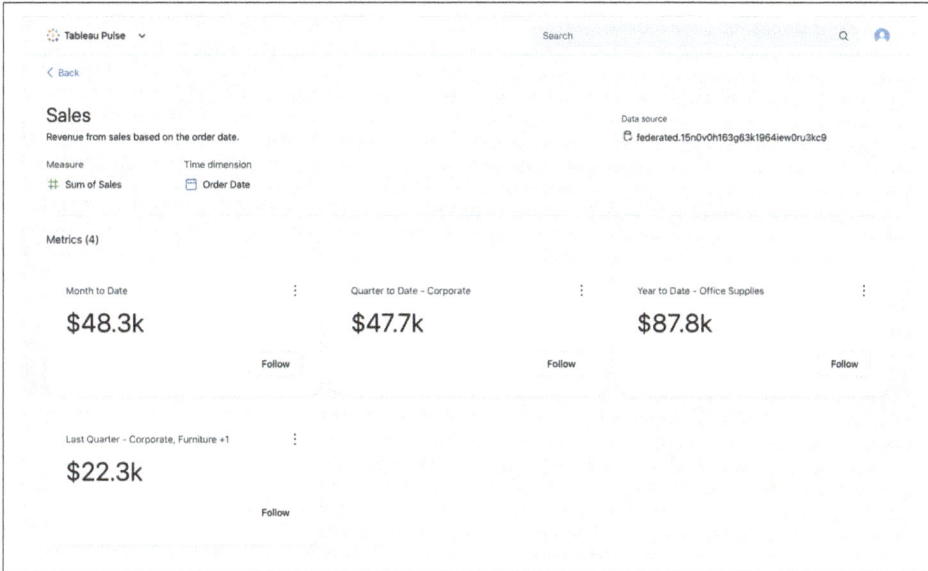

Figure 4-2. Metrics available for the Sales metric definition (see a larger version of this figure online (https://oreil.ly/lait0402))

Creating New Metrics

The metric creation process for an end user is identical to what you saw in Chapter 2. To construct a metric, users can either access the metric definition by using the Insights Exploration section or, more likely, they can begin by clicking a specific metric that they may want to customize further. It is important to note that any new metrics they create need to have at least one follower assigned before leaving the screen. This will ensure that the metric is created and accessible for others.

To see how the AI-generated insight summaries are propagated to end users, you can create your own metrics or use metrics outlined in the following sidebar:

- Year to Date Profit Ratio
- Year to Date Average Discount
- Year to Date Number of Customers
- Year to Date Items Sold
- Year to Date Profit

- Year to Date Average Time to Ship
- Year to Date Average Products per Order
- Year to Date Average Profit
- Year to Date Orders Shipped

Additional Metrics for Superstore Transactions

You may want additional metrics available to further test Tableau Pulse functionality. The Superstore Transactions data set includes many options for constructing such metrics. Table 4-1 shows the metric definitions with suggested filters, number formatting, and more. Those marked with an asterisk are advanced definitions.

Table 4-1. Additional metrics using the Superstore Transactions data set

Metric name	Measure and aggregation	Time dimension	Adjustable filters	Number format	Value going up is
Orders Shipped	Distinct Count of Order ID	Order Date	Ship Mode, Region	Number	Favorable
Average Profit	Average of Profit	Order Date	Segment, Region	Currency	Favorable
Average Products per Order*	`AVG({FIXED Order ID: COUNTD(Product ID)})`	Order Date	Category, Segment	Number	Favorable
Average Time to Ship*	`AVG({FIXED Order ID: DATEDIFF('day',Order Date,Ship Date)})`	Order Date	Ship Mode, State/Province	Number; day, days	Unfavorable
Profit	Sum of Profit	Order Date	Sub-Category, Product Name	Currency	Favorable
Items Sold	Sum of Quantity	Order Date	Category, Sub-Category	Number; item, items	Favorable
Number of Customers	Distinct Count of Customer ID	Order Date	Region, Segment	Number; customer, customers	Favorable
Average Discount	Average of Discount	Order Date	Segment, State/Province	Percentage	Unfavorable
Profit Ratio*	`SUM(Profit)/ SUM(Sales)`	Order Date	Category, Sub-Category	Percentage	Favorable

Figure 4-3 shows the nine metrics from the sidebar after following them. Followed metrics are organized in chronological following order. Remember, Pulse is heavily dependent on today's date, so the generated summaries and insights may differ significantly in your environment.

> Tableau Cloud users with a Tableau+ license can sort metrics in ascending or descending order by data source name, metric definition, or recently followed. A static goal value can be applied to metrics with a period-to-date time range. The goal will display underneath the large number with a progress bar and within the line chart as a reference line if the sparkline values are set to show as a running total. You can more about Tableau+ in Chapter 7.

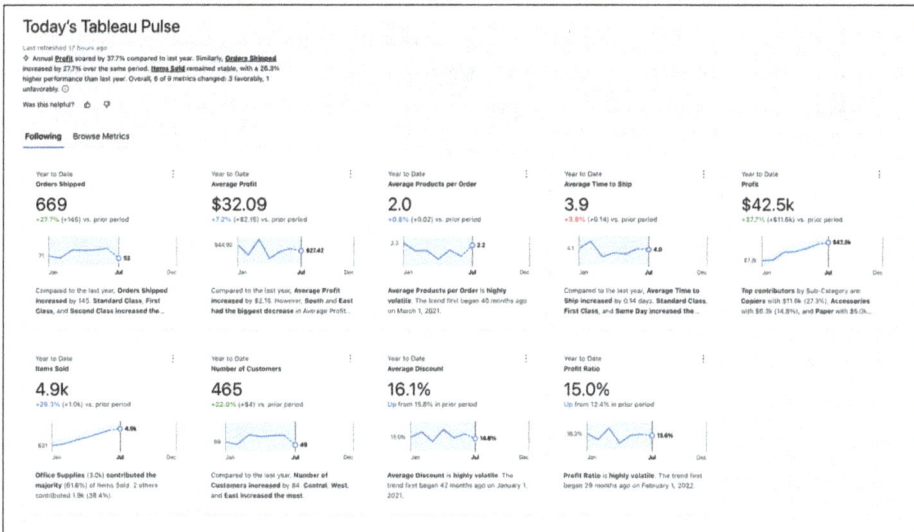

Figure 4-3. Pulse summary after following metrics (see a larger version of this figure online (https://oreil.ly/lait0403))

Depending on what happens when metrics are followed, Pulse will begin to construct a larger insight summary that combines the most interesting insights from the user's followed metrics. These summaries are constructed every 24 hours and may take at least a day to initially show up for the end user. The following is the AI summary readout from Figure 4-3. Metrics in this summary show up as bold and underlined, and the user can click them to navigate directly to the detailed view of the metric:

Annual **Profit** soared by 37.7% compared to last year. Similarly, **Orders Shipped** increased by 27.7% over the same period. **Items Sold** remained stable, with a 26.3% higher performance than last year. Overall, 6 of 9 metrics changed: 3 favorably, 1 unfavorably.

Understanding Generated Insights

As mentioned in Chapter 2, Tableau Pulse relies on the dimension fields specified for adjustable filters to generate insights. Behind the scenes, insight types (refer to Table 2-3) are generated and then scored and ranked to determine which are most statistically significant. These insights appear in both the subsequent Breakdown charts and the automated insight summaries.

End users have the option to provide feedback on the helpfulness of each insight by way of a thumbs-up or thumbs-down button next to each generated insight. When the insight summary is tagged as *helpful*, users can provide a description of up to 3,000 characters. Similarly, as seen in Figure 4-4, users can tag an insight summary as *not helpful*, select reasons, and also provide feedback. The options for this type of feedback are as follows:

Biased, toxic, or harmful
    Language in the output is offensive in some way.

Inaccurate
    The numerical values generated in the output are factually wrong.

Incomplete
    The output doesn't draw a conclusion.

Inappropriate tone or style
    Language in the output isn't professional or is overly colorful.

Other
    This is for all other issues with the generated output.

Figure 4-4. Types of feedback that can be selected for insight summaries

This mechanism of feedback doesn't provide any immediate change to the insights generated. Instead, it is passed to the LLM responsible for generating the summaries (likely by way of prompt instructions).

> There is currently no way to *undo* marking a generated insight or chart as helpful or not helpful. Since this can have an impact on subsequent behavior, I recommend caution when assigning feedback.

Metric Digests

In addition to the Pulse home screen of followed metrics, users can receive digests of their followed metrics. The default setting for these digests is weekly emails and Slack messages, but the frequency can be adjusted to daily, and users can specify which communication channels they want to receive the summary in. Figure 4-5 shows the options available, accessed by clicking the user icon in the upper right of the Pulse home screen and selecting Preferences. Slack must first be configured within a Tableau Cloud site by an administrator to be available as a channel option. You'll find out how to configure Tableau Cloud and Slack in Chapter 6.

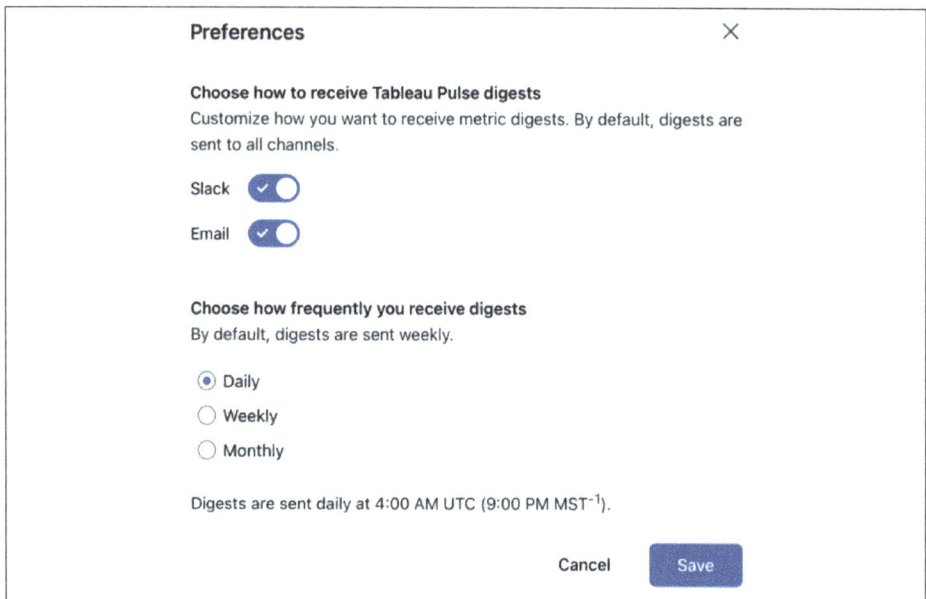

Figure 4-5. Preferences that end users can set for Pulse digests

These digests coincide with updates to the generated AI summaries and cannot be configured to a specific time. Based on observation, it appears that the daily insight summaries are constructed somewhere between 3 a.m. and 4 a.m. Coordinated Universal Time (UTC), and then the digests are subsequently sent out.

Email Digest

Within the email digest, you'll see the same generated summary of the metrics, stating which metrics have experienced the most interesting changes. Beneath the global summary are thumbnails, metric specific summaries, and hyperlinks to view the details of each metric. Metrics in this format are displayed in alphabetical order. Figure 4-6 shows a fragment of a daily email digest.

Tableau Pulse

Hello, **Tableau Viewer!**
Here are your Tableau Pulse metrics

◇ Annual **Profit** soared by 37.7% compared to last year. Similarly, **Orders Shipped** increased by 27.7% over the same period. **Items Sold** remained stable, with a 26.3% higher performance than last year. Overall, 6 of 9 metrics changed: 3 favorably, 1 unfavorably.

This tool uses generative AI, which can produce inaccurate or harmful responses.

Send feedback from Tableau Pulse

Year to date: Jan 1–Jul 15, 2024
Average Discount

16.1%

No change vs. prior year

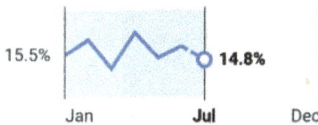

Average Discount is highly volatile. The trend first began 42 months ago on January 1, 2021.

Show Details ⌴

Year to date: Jan 1–Jul 15, 2024
Average Products per Order

2.0

No change vs. prior year

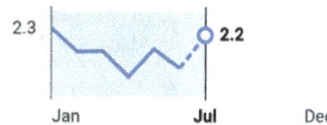

Average Products per Order is **highly volatile**. The trend first began 40 months ago on March 1, 2021.

Show Details ⌴

Figure 4-6. A snippet of an email digest (see a larger version of this figure online (https://oreil.ly/lait0406))

The email also contains footer links for the email recipient to manage the metrics they follow or to adjust their digest frequency settings.

Slack Digest

Users can also enable Slack digests, which allow them to see the same Pulse insights directly in their communication platform. Once a Tableau Cloud administrator enables the integration between Slack and Tableau Cloud via the Tableau for Slack app, each user must add the Tableau app to their Slack environment and authorize the connection between the two applications.

The Tableau app can be added by clicking "Add apps" in the workspace sidebar within the Slack workspace, as shown in Figure 4-7.

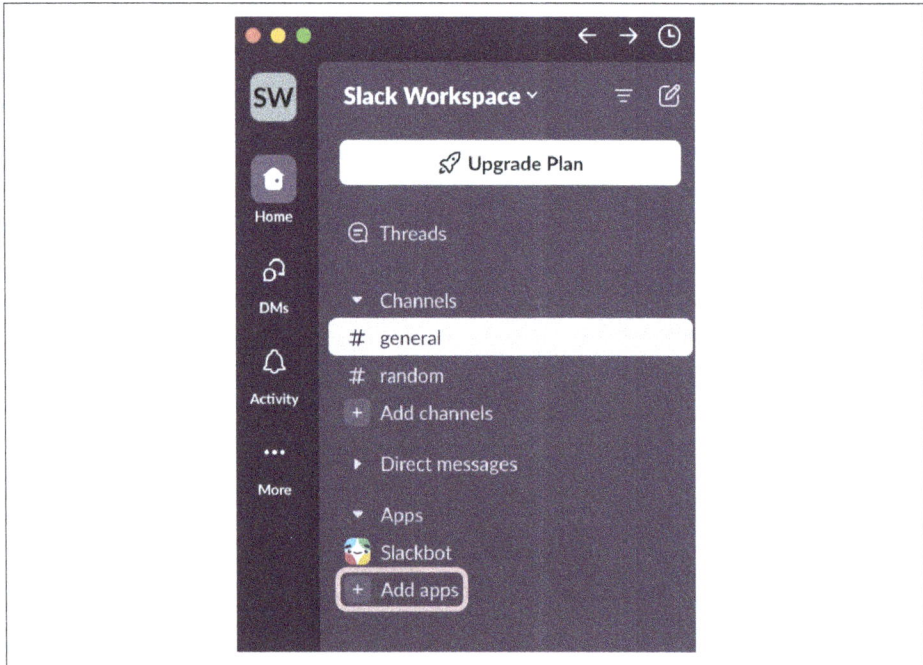

Figure 4-7. Slack workspace sidebar with the "Add apps" option highlighted

Users will then be able to select the Tableau app from the list of available apps and begin the authorization process with Tableau Cloud. Authorization requires users to sign in to Tableau Cloud and allow Slack to perform actions on behalf of the user in Tableau Cloud. Figure 4-8 shows the authorization screen with the explicit list of information and actions Slack will be granted.

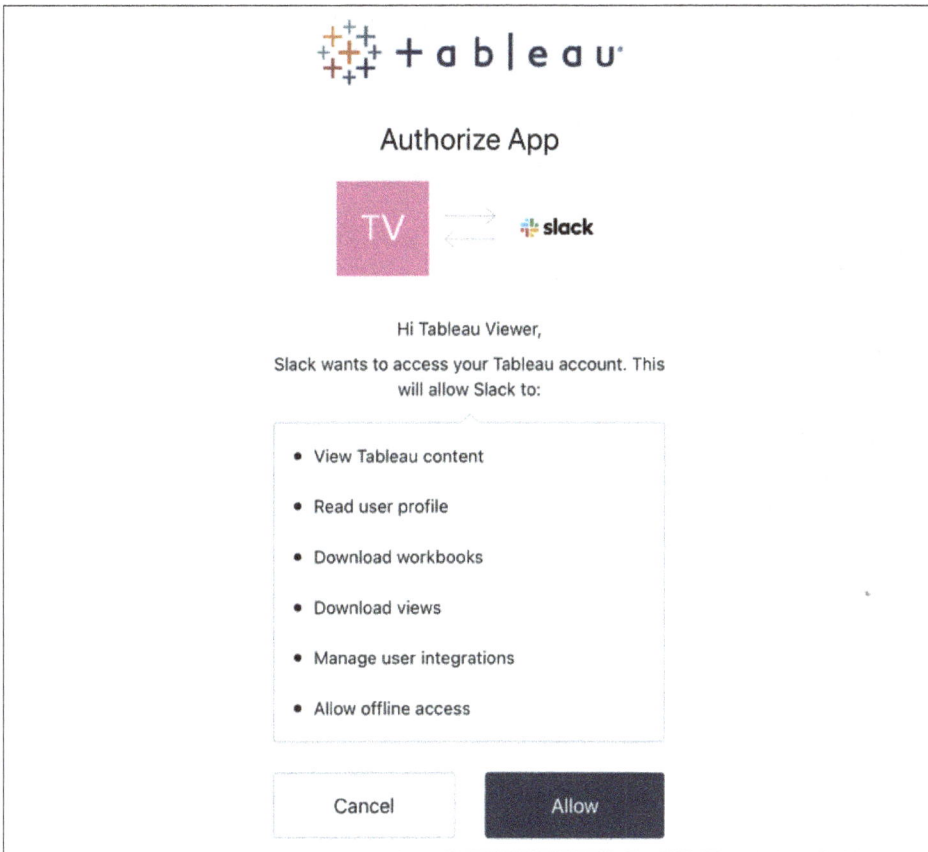

Figure 4-8. List of authorizations given to Slack when adding the Tableau app

> If the Tableau app isn't already available to users in your Slack workspace and you don't have permissions to add it, you can initiate a request to your Slack administrator by following the same process of attempting to add the app. You'll end up in an approval request screen with the option to include more information about your request. Slackbot will direct message (DM) you once your request is reviewed by an administrator.

Once the Tableau app is enabled for a user, they will begin to receive Pulse digests at their desired frequency. These come in the form of DMs from the Tableau app in Slack. These digests mirror what is seen in the email digests, starting with the general summary of all metrics followed. Each followed metric is re-created with its current value, comparison value(s), sparkline, and insights. Similar to the email digest, metrics here are shown in alphabetical order. If a user follows many metrics (more than four), the Tableau app will include remaining metrics in a threaded reply. Figure 4-9 shows an example digest.

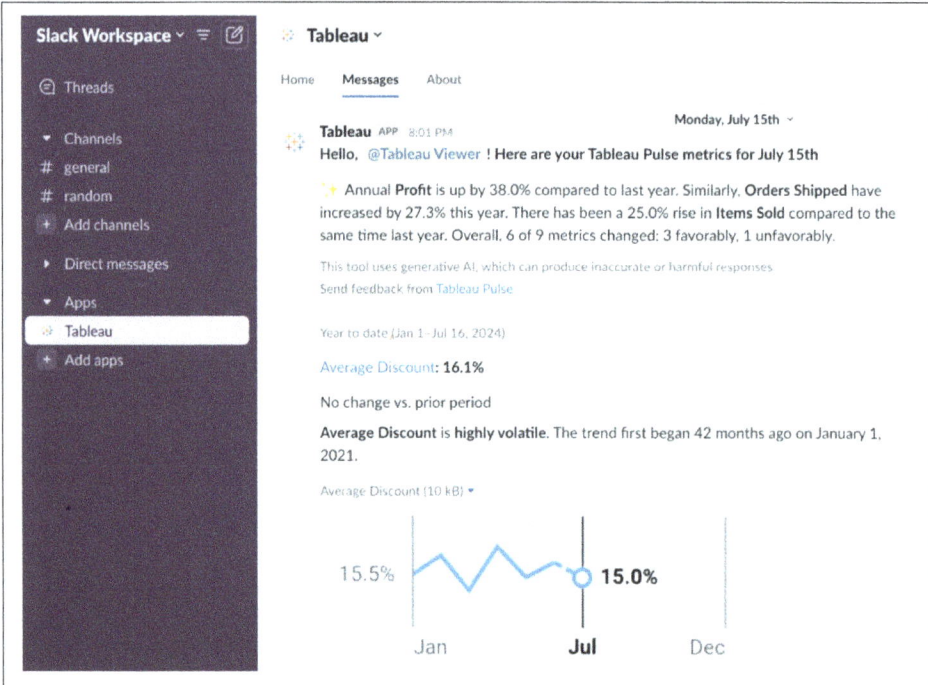

Figure 4-9. Tableau Pulse digest in Slack (see a larger version of this figure online (https://oreil.ly/lait0409))

Metrics within these digests have hyperlinks to directly navigate to the insight exploration screen. Management buttons, which are identical to the email digest options at the bottom of the message, allow the user to manage their metrics or digest preferences.

Tableau Mobile

Users can also access Pulse metrics and summaries by using the Tableau Mobile app. This app is available for free on both Android and iOS, giving users the ability to access their Tableau content on the go. Figure 4-10 shows what Tableau Pulse looks when viewed from an iPhone.

Within the Tableau Mobile app, users can scroll vertically through their followed metrics (which are displayed in chronological follow order, like on Tableau Cloud). In this regard, the Mobile version is identical to the digests. The user can also interact with the detailed view of the metric and both ask and access top insights. If notifications are configured, the user can receive reminders to check their Pulse insights. Users can also set Pulse as their starting page when opening the app.

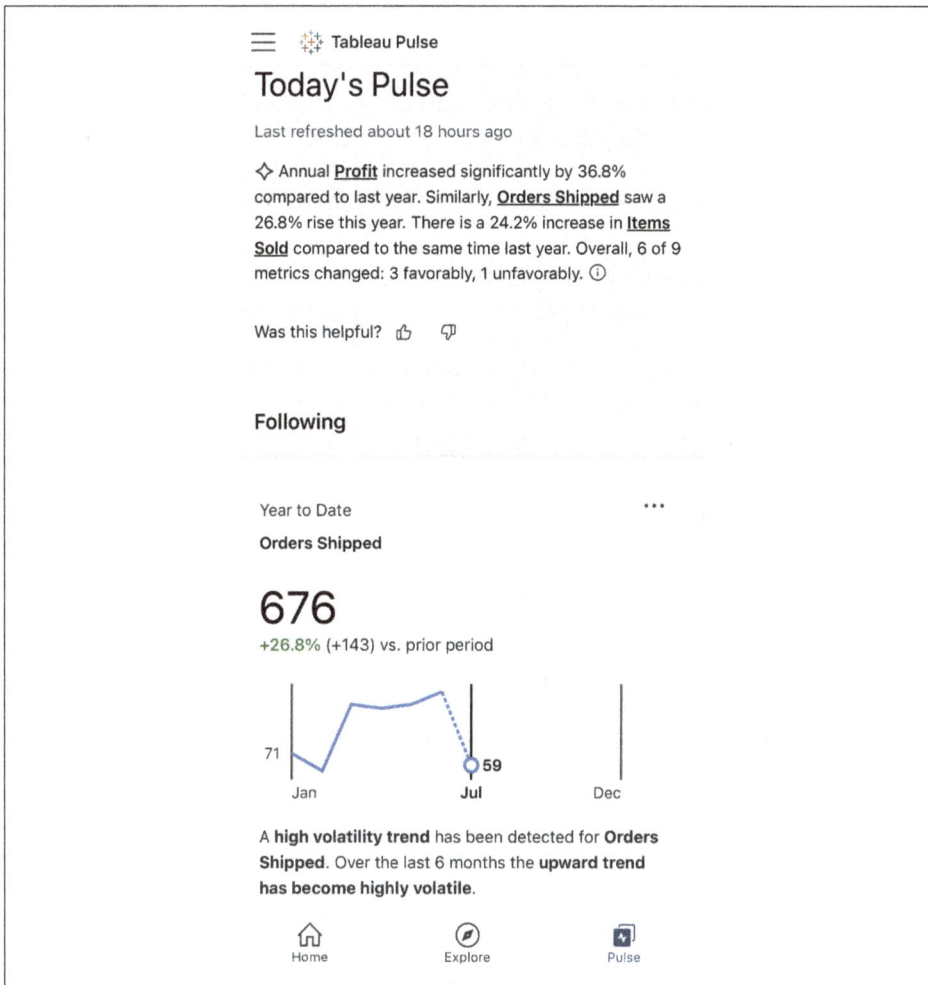

Figure 4-10. Tableau Pulse summary in Tableau Mobile (see a larger version of this figure online (https://oreil.ly/lait0410))

It is important to note that unlike with the browser experience, users are unable to create new metrics on their phone. Additionally, they won't have access to the adjustable filters tied to each metric other than the time horizon.

Comparing the different methods of accessing the digest, the Mobile app experience offers the richest and easiest-to-use experience. Users can seamlessly toggle between the summary and detailed views of their followed metrics, the metrics themselves are larger and easier to read, and all the visualizations are interactive. For busy executives, it's the easiest way to access all their metrics in one place.

Summary

This chapter was an in-depth look at what end users experience when using Tableau Pulse. Here are some key takeaways for what you've learned:

- Users with Viewer licenses can construct metrics based only on premade metric definitions.
- All users, regardless of license type, can manage and modify followers of metrics.
- Users can tag insights as *helpful* or *not helpful* and provide feedback.
- Users can receive digests via email or Slack on a configurable schedule (daily, weekly, or monthly).
- Both email and Slack digests include hyperlinks to access and manage followed metrics.
- A user must enable the Tableau app for Slack and authorize Slack to begin receiving digests.
- The Tableau Mobile app includes Tableau Pulse and provides the easiest and most immersive experience for accessing followed metrics.
- Users cannot create new metrics or access adjustable filters in the Mobile app.

In Chapter 5, you will see how a variety of industries can benefit from using Tableau Pulse. The chapter will discuss several metric examples and expose you to a variety of data sources. You'll also see metrics that can be created from the Admin Insights data sources mentioned in Chapter 3, giving you tools to monitor usage and adoption of Pulse across your Tableau site.

Use Cases with Tableau Pulse

In the previous chapters, you've seen all the building blocks it takes to construct metrics in Tableau Pulse. Now it is time to see what metrics in action look like for different industries. In this chapter, I share use cases for sales, finance, healthcare, and administering Tableau Pulse itself. My goal is that you'll walk away from this chapter inspired by how you can implement Tableau Pulse into your analytics environment and how your end users can benefit from Pulse. In addition to giving you some inspiration, I'll walk you through how to maximize and customize the insights section of Pulse metrics.

Use Case: Sales

Sales data is one of the most popular types of data sets you're likely to work with. Rare is the company that doesn't have some type of product that they are offering up to customers. So in this first use case, I want to show you how to re-create an existing Tableau dashboard focused on a pipeline of sales opportunities for a software company using Pulse. This exercise can be helpful if you're tasked with transitioning existing content or if you need to make a decision whether to start with Pulse or to construct an entire dashboard. It will also help illustrate the type of insights that are automatically generated versus what your end users may have to determine on their own.

Figure 5-1 shows a dashboard where an end user can view several key performance indicators (KPIs), and using the dropdown in the upper-right corner, can toggle between three time periods: MONTH TO DATE, QUARTER TO DATE, and YEAR TO DATE. Additionally, the user can click either of the two bottom charts to interactively filter by the opportunity type, the stage the opportunity is in, and the territory managing the opportunity.

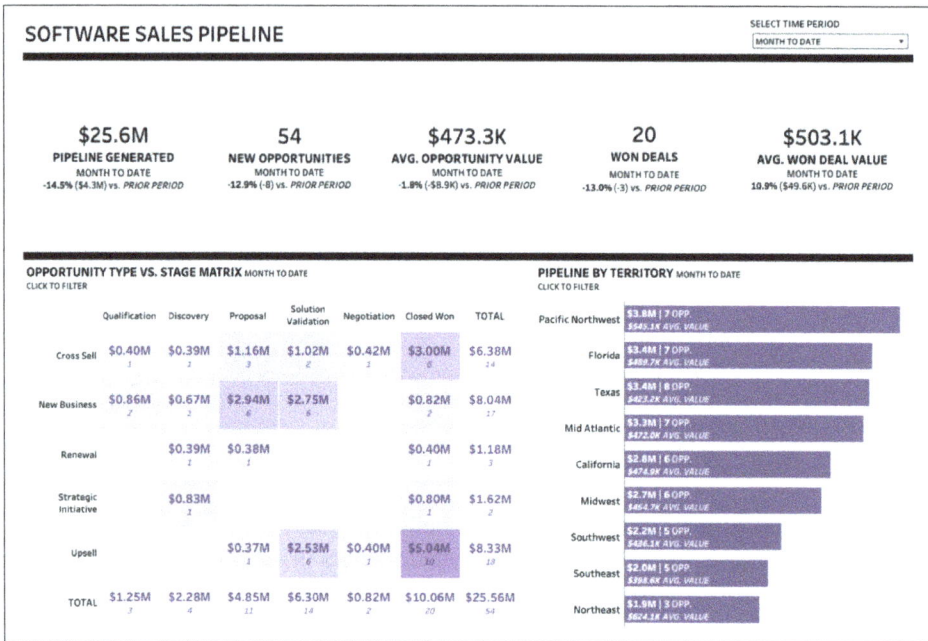

Figure 5-1. A software sales pipeline dashboard (see a larger version of this figure online (https://oreil.ly/lait0501))

The layout of the dashboard in Figure 5-1 is very popular in the BI world. It includes high-level KPIs and the ability to change time periods for end of month/quarter/year reporting. Additionally, it has some detail on the bottom half, quickly displaying the distribution of the opportunities by important categories. The inclusion of comparison to prior time periods provides the context end users crave to determine whether they're on target to meet their goals.

Now, let's begin unpacking and re-creating this dashboard in Pulse. First, I'll address one of the KPIs: how it is made in Tableau Desktop and how it can be made in Pulse. Figure 5-2 shows the worksheet for the KPI, with the key areas driving the visualization highlighted.

Three key sections are driving this visualization:

Marks card
> Highlighted on the Marks card, three calculations have been constructed to generate the pipeline value for the time period chosen. These dynamic calculations change based on user selection.

TIME PERIOD (bottom of figure)

Below the Marks card, this has a drop-down selection set to MONTH TO DATE. This is the controller that allows users to select time periods.

Filters shelf (top of figure)

Above the Marks card, this shows the three filters in place. There are two Action filters, which filter the worksheet based on Sales Territory and Opportunity Stage/Opportunity Type. These correspond to the two charts at the bottom of Figure 5-1. There is also a simple exclude filter (Won Flag: False) to remove any lost opportunities, as those don't contribute to the pipeline.

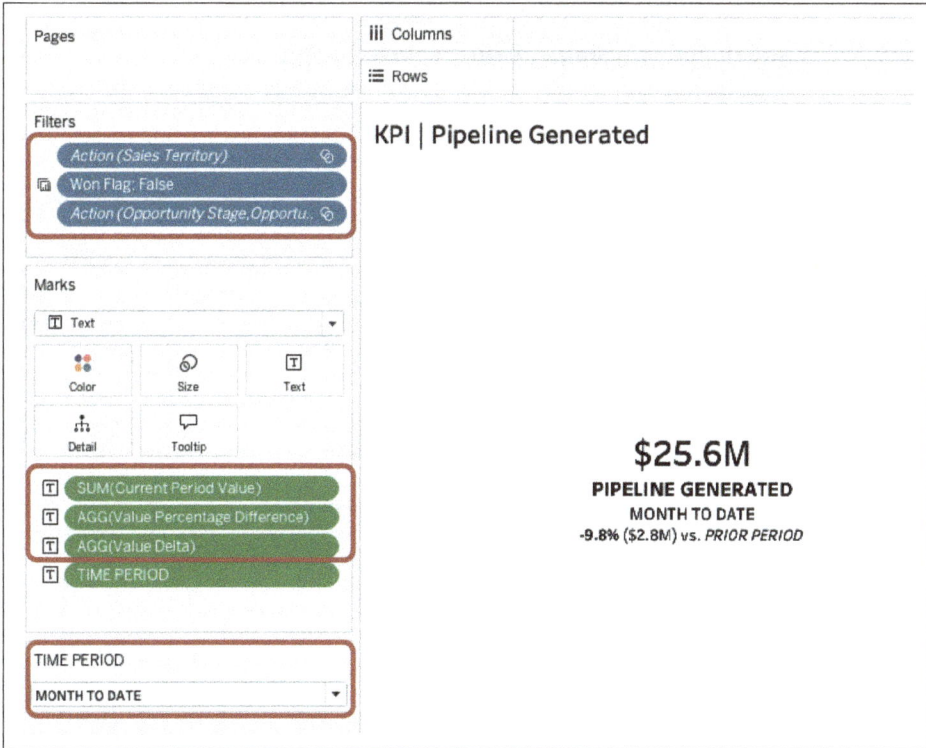

Pages	iii Columns	
	≣ Rows	

Filters
- Action (Sales Territory)
- Won Flag: False
- Action (Opportunity Stage, Opportu...

KPI | Pipeline Generated

Marks

⊤ Text ▾

Color   Size   Text

Detail   Tooltip

- ⊤ SUM(Current Period Value)
- ⊤ AGG(Value Percentage Difference)
- ⊤ AGG(Value Delta)
- ⊤ TIME PERIOD

$25.6M
PIPELINE GENERATED
MONTH TO DATE
-9.8% ($2.8M) vs. *PRIOR PERIOD*

TIME PERIOD
MONTH TO DATE ▾

Figure 5-2. Worksheet for a KPI

The parameter and calculations are the secret sauce that make this KPI work, but for someone newer to the Tableau world, they aren't exactly easy or intuitive to make. Even for a seasoned professional, they are very calculation-intensive, requiring three calculations for each KPI being visualized (and several more that those calculations are dependent on).

Here's what's actually going on within the calculations:

- Six Boolean calculations isolate different time periods, both the *current* time period and the *prior* time period. They rely on today's date to determine whether an opportunity is within either of the time periods and will be used to filter data in subsequent calculations.

- There are two case statement calculations, one for each of the time periods, to determine which Boolean should be used for the filtering calculations.

- There are four calculations, one for each of the time period values, a percentage difference, and a difference (delta) between the current and prior period values.

Figure 5-3 shows all the calculations constructed for the KPI.

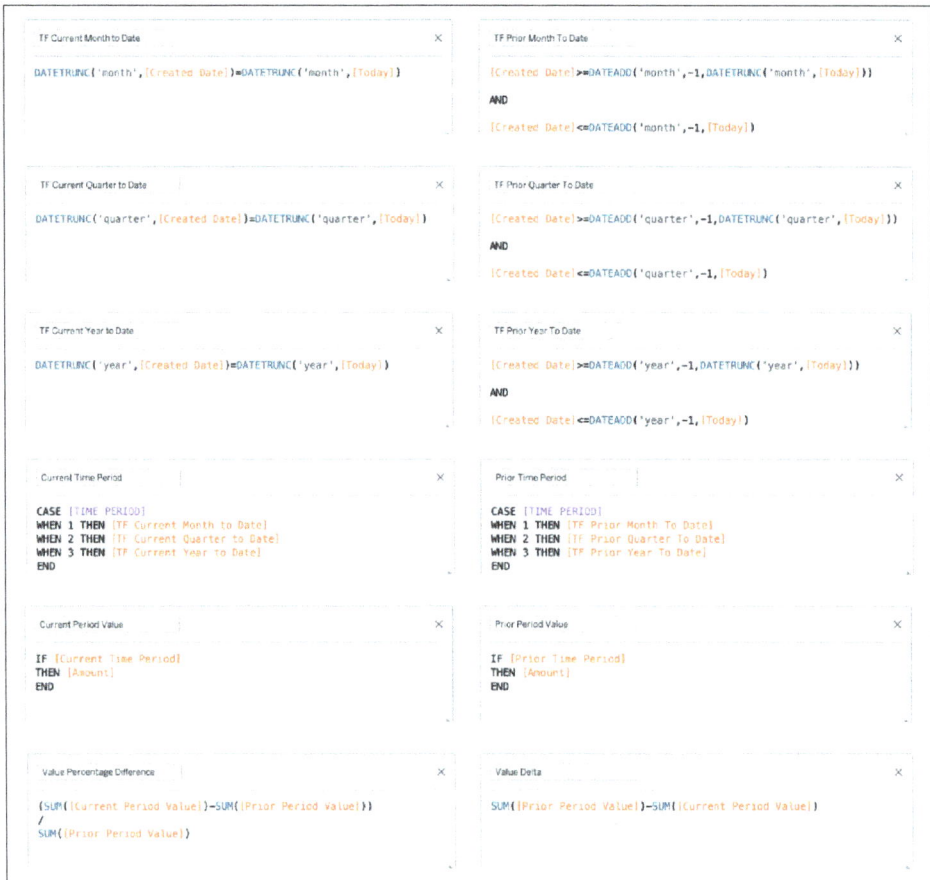

Figure 5-3. The calculations required to construct a KPI (see a larger version of this figure online (https://oreil.ly/lait0503))

Now that you've seen what's required to create that seemingly simple KPI in Tableau Desktop, I'm going to re-create it as a Pulse metric. By now you're an expert at creating metrics in Pulse, so it should be no surprise that this is an easy task. The steps are straightforward; I define the following:

- The Measure used is Amount with an Aggregation of Sum.
- The "Definition filters" dropdown is set to exclude false values for the Won Flag.
- The "Time dimension" is Created Date.
- Only "Prior period" is selected for Time Comparison; the Secondary comparison is set to None.
- In Options, to allow for the same filtering experience as the interactive filtering, three fields are added as "Adjustable metric filters": Sales Territory, Opportunity Type, and Opportunity Stage.
- "Number format" is set to Currency > USD.
- For the Insights type in Dimensions, "Value going up is" is set to Favorable, and Record-Level Outliers are turned off.

Figure 5-4 shows all the metric definition configurations in place, which truthfully took longer to screenshot than to select in the dropdowns.

Right away you should be seeing the benefit that Pulse brings to visualizations involving comparison time periods. Instead of constructing calculations for each unique time period to be compared, it's done automatically, and there are many more valuable time comparisons the dashboard didn't feature. Similarly, the necessity to construct the percentage difference and value difference is no longer required, and instead these are automatically constructed.

Now to get down to the actual metric and spend some time assessing the automatically generated and summarized insights. This will help you understand the type of information those consuming Pulse metrics will be exposed to. Figure 5-5 shows the metric created based on the configurations previously specified.

On the metric card, there's a summary listing the top contributors by Opportunity Type, which was one of the selected adjustable metric filters. It is a valuable sound bite, and the inclusion of the percentages along with the actual values adds context to understand the distribution. Within the breakdown, you can toggle between the three categories to quickly compare the distribution of opportunities.

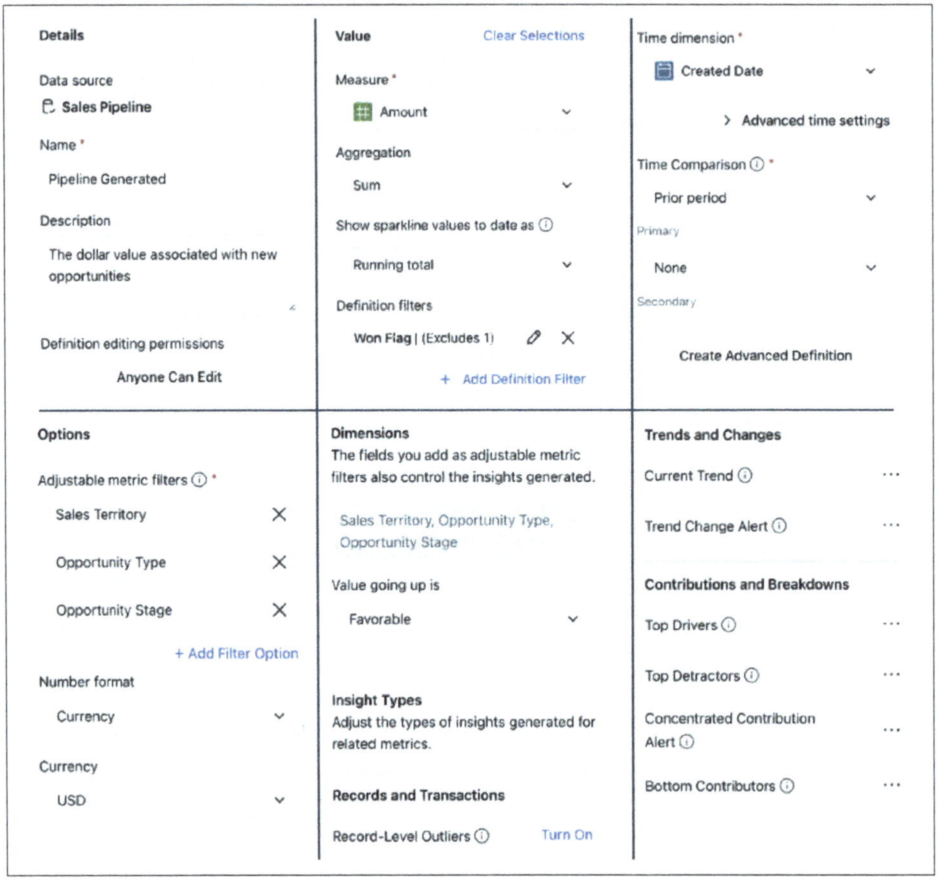

Figure 5-4. Configurations for the Pipeline Generated KPI in Pulse (see a larger version of this figure online (https://oreil.ly/lait0504))

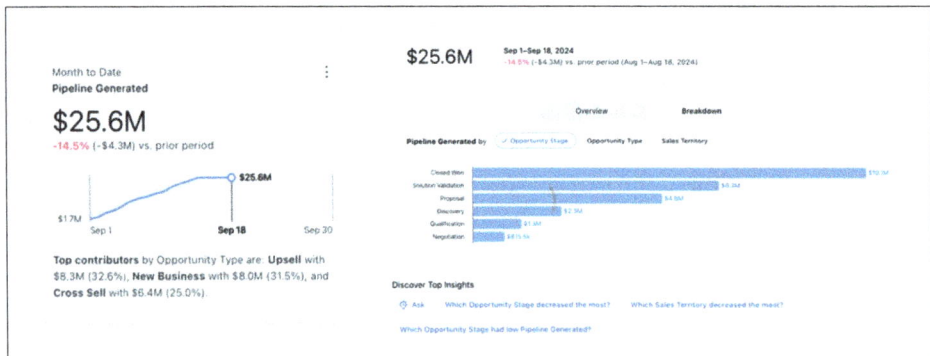

Figure 5-5. Metric card and breakdown (see a larger version of this figure online (https://oreil.ly/lait0505))

Additionally, 12 pregenerated questions are in the Discover Top Insights section, each with its own companion bar chart and summarized statement. Here are the 12 questions:

- Which Sales Territory increased the most?
- Which Sales Territory decreased the most?
- Which Sales Territory had low Pipeline Generated?
- Which Sales Territory had high Pipeline Generated?
- Which Opportunity Type increased the most?
- Which Opportunity Type decreased the most?
- Which Opportunity Type had high Pipeline Generated?
- Which Opportunity Stage had high Pipeline Generated?
- Which Opportunity Stage had very high Pipeline Generated?
- Which Opportunity Stage had low Pipeline Generated?
- Which Opportunity Stage increased the most?
- Which Opportunity Stage decreased the most?

Of the questions generated, those aimed at the high and low pipeline generated are immediately visible both in the breakdown section of the metric and the dashboard the metric was based on. However, the remaining questions and their answers are harder to surface within the dashboard and do present interesting information. As an example, Figure 5-6 shows the chart and insight for the first question, aimed at demonstrating which Sales Territory increased its pipeline the most over the time period, a task that would take multiple clicks within the dashboard to uncover the same information. Serving it up automatically with a summary sentence and companion chart fully demonstrates the time savings that users may gain through Pulse.

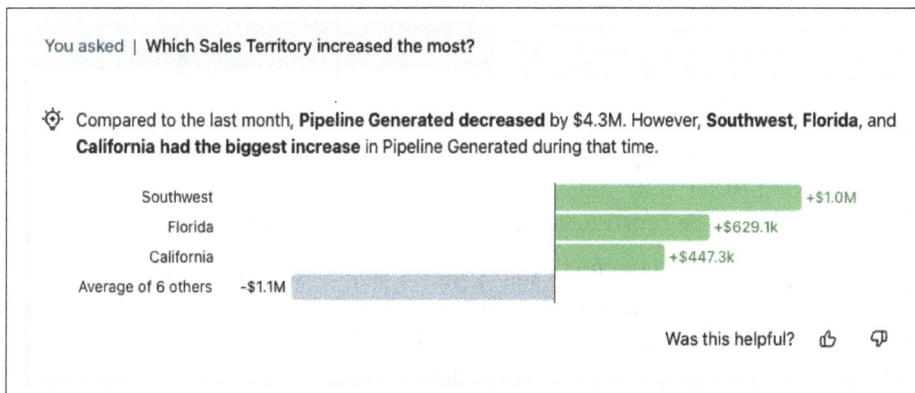

Figure 5-6. A top insight generated by Pulse

Along the same lines, you'll notice that none of the insights generated feature a combination or mixture of the fields (Sales Territory, Opportunity Type, Opportunity Stage) available for filtering or selection within the breakdown. So, potential insights like "Which Opportunity Stage are most New Business opportunities in?" that relates the intersection of Opportunity Stage and Opportunity Type aren't immediately visible in Pulse, unlike the heat map on the dashboard. To get to this level of insight, an additional child metric from the existing metric definition, with the Opportunity Type set to New Business, will be necessary, as shown in Figure 5-7.

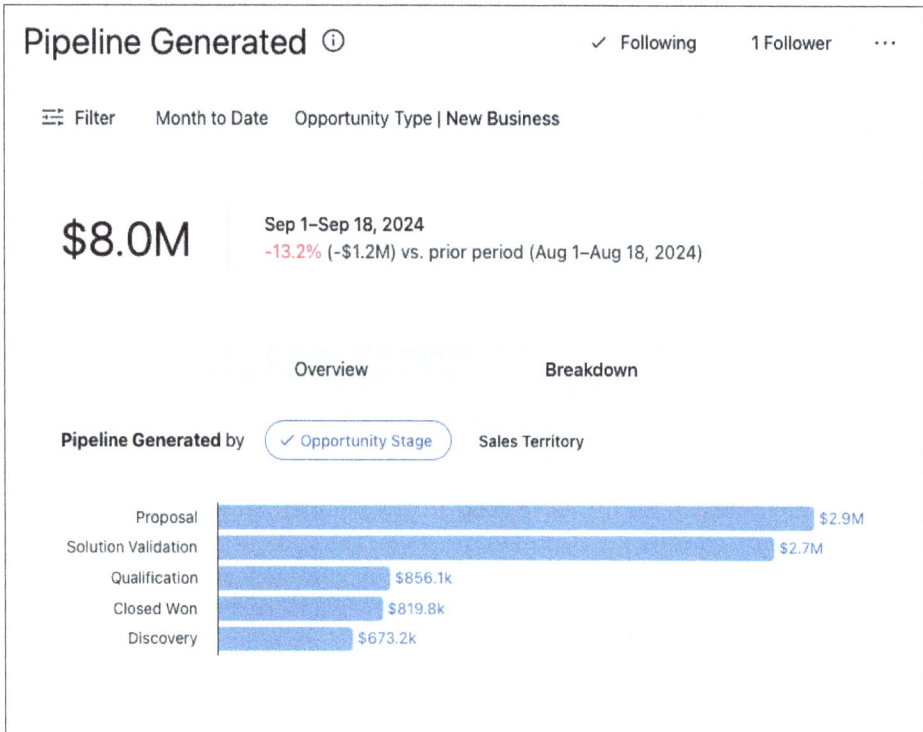

Figure 5-7. A child metric with Opportunity Type filtered (see a larger version of this figure online (https://oreil.ly/lait0507))

Knowing the types and depth of insights generated can help guide you as to whether Pulse is a good fit for your audience. It also provides direction on how many child metrics may need to be constructed to answer more pointed questions.

Finally, Figure 5-8 shows the Pulse start page for someone who has followed the five metrics featured in the dashboard, fully demonstrating what the Pulse experience would be like. Certainly the short, summarized insights clearly offer a different experience than the self-service and interactive discovery experience of the dashboard.

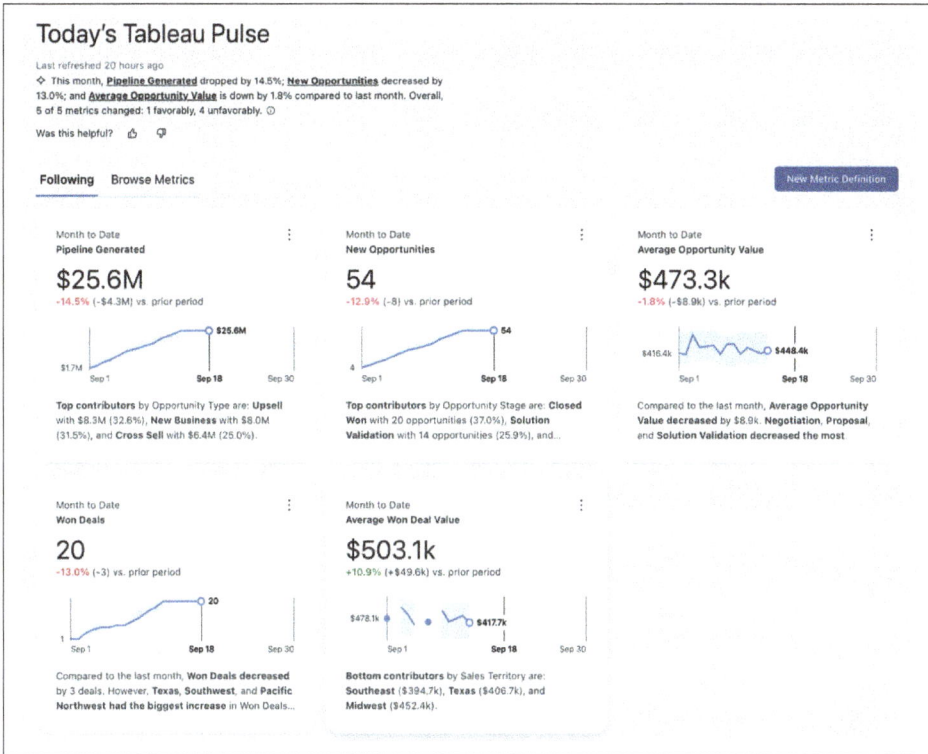

Figure 5-8. KPIs from the dashboard as Pulse metrics (see a larger version of this figure online (https://oreil.ly/lait0508))

Use Case: Finance

With Tableau Pulse, the speed at which insights can be constructed offers a significant advantage over building complex calculations within a dashboard. The rapidity with which Pulse can enable data exploration can be a game changer, especially when the types of automated insights are customized. In this use case, I'm going to show you how a simple data set of credit card transactions can quickly add value and also enable anomaly detection in real time.

To add some context to this use case, imagine this project is for a small regional credit union that offers a high level of customer service. The credit union appeals to affluent customers who are also busy professionals. As part of the digital transformation the organization is undergoing, the product manager for mobile payments wants to understand customer habits and keep an eye out for any risky behaviors.

Figure 5-9 shows a snippet of the transaction data set. Currently the product manager has access to a reporting system where she can export transaction information, but she doesn't have any real-time analytics or method to track the information she's interested in. Let's explore how Tableau Pulse can support her need for information.

Account ID	Account Holder	Date	Transaction Type	Transaction ID	Merchant Code	Merchant Name	Amount	Transaction Detail
56416262	Kare Brockhurst	1/1/2022	Swipe	1890120820350743	5541	Service Stations	9452.62	5XX8262|SWIPE|5541-SERVICESTAT|TXX5074|9452.62USD
54556613	Constantino Pam	1/1/2022	Apple Pay	5742219841747571	4814	Telecommunications Services	122.95	5XX6613|APPLEPAY|4814-TELECOMMUNIC|TXX4757|122.95USD
55644441	Judy Halpine	1/1/2022	Contactless	6649797679186918	5411	Grocery Stores	192.3	5XX4441|CONTACTLESS|5411-GROCERYSTOR|TXX8691|192.3USD
59545532	Bloise Gwilt	1/1/2022	Apple Pay	6874442625125928	4814	Telecommunications Services	360.44	5XX5532|APPLEPAY|4814-TELECOMMUNIC|TXX2592|360.44USD
54342640	Isa Twentyman	1/1/2022	Google Pay	2805708623899077	6536	Money Transfer Services	66.7	5XX2640|GOOGLEPAY|6536-MONEYTRANSF|TXX9907|7|66.70USD
57767924	Corbie MacAnelly	1/1/2022	Apple Pay	8179698598384439	5541	Service Stations	227.69	5XX7924|APPLEPAY|5541-SERVICESTAT|TXX8443|9|227.69USD
58472948	Aloysius Telwood	1/1/2022	Contactless	9476749384465950	5411	Grocery Stores	479.2	5XX2948|CONTACTLESS|5411-GROCERYSTOR|TXX8595|0|479.20USD
53408936	Kizzee de Voiels	1/1/2022	Swipe	2533270536291805	6536	Money Transfer Services	243.81	5XX8936|SWIPE|6536-MONEYTRANSF|TXX1805|243.81USD
58561354	Skally Silverston	1/1/2022	Chip Insert	3191456544262741	5812	Eating Places/Restaurants	300.95	5XX1354|CHIPINSERT|5812-EATINGPLACE|TXX82741|300.95USD
81370986	Blinni Sharpless	1/1/2022	Apple Pay	7352499324336154	6536	Money Transfer Services	142.71	5XX0966|APPLEPAY|6536-MONEYTRANSF|TXX36154|142.71USD

Figure 5-9. A subset of credit card transactions

First, it's worth mentioning that this data set and the task at hand are an optimal candidate for Tableau Pulse. The data itself is clean and well structured, and the requests to understand customer behavior and assess risks in real time lend themselves to the up-to-date insights that Pulse generates.

Figure 5-10 shows the various Pulse metrics constructed to help out the product manager. I'm going to unpack them at a high level and then focus on one specific metric and how it's been customized to tackle her request to identify any risky behaviors.

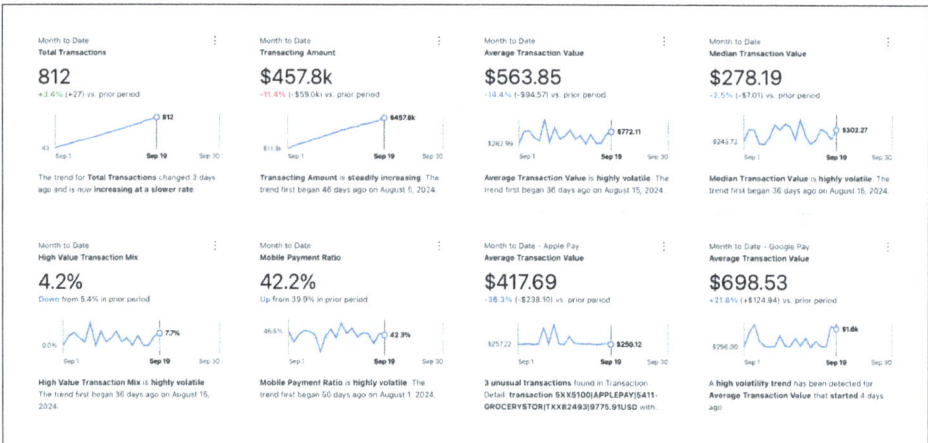

Figure 5-10. Pulse metrics based on credit card transactions (see a larger version of this figure online (https://oreil.ly/lait0510))

The metrics the product manager now has access to consist of a combination of analytical scopes. The top row contains operational metrics, providing insight into overall trends and behaviors at the credit union. The bottom row includes more specific metrics related to her needs for understanding risk and mobile payment behavior. This mix of metrics provides a great launching-off point, giving her the ability to understand the organization's overall customer behaviors alongside her focus area.

There's immediate value gained with the ability to compare the Average Transaction Value for customers against those that are specific to Apple Pay and Google Pay. Similarly, she can use the Mobile Payment Ratio to understand whether her team's efforts are increasing usage of mobile payment methods.

Beyond the quick insights she's gained, you can see something interesting happening in the Average Transaction Value metric that's filtered to include only Apple Pay transactions. Shown in Figure 5-11, there appear to be three unusual transactions of this payment type. Let's unpack the details of this metric to see not only what it has found, but how the metric has been tuned to find the unusual transactions.

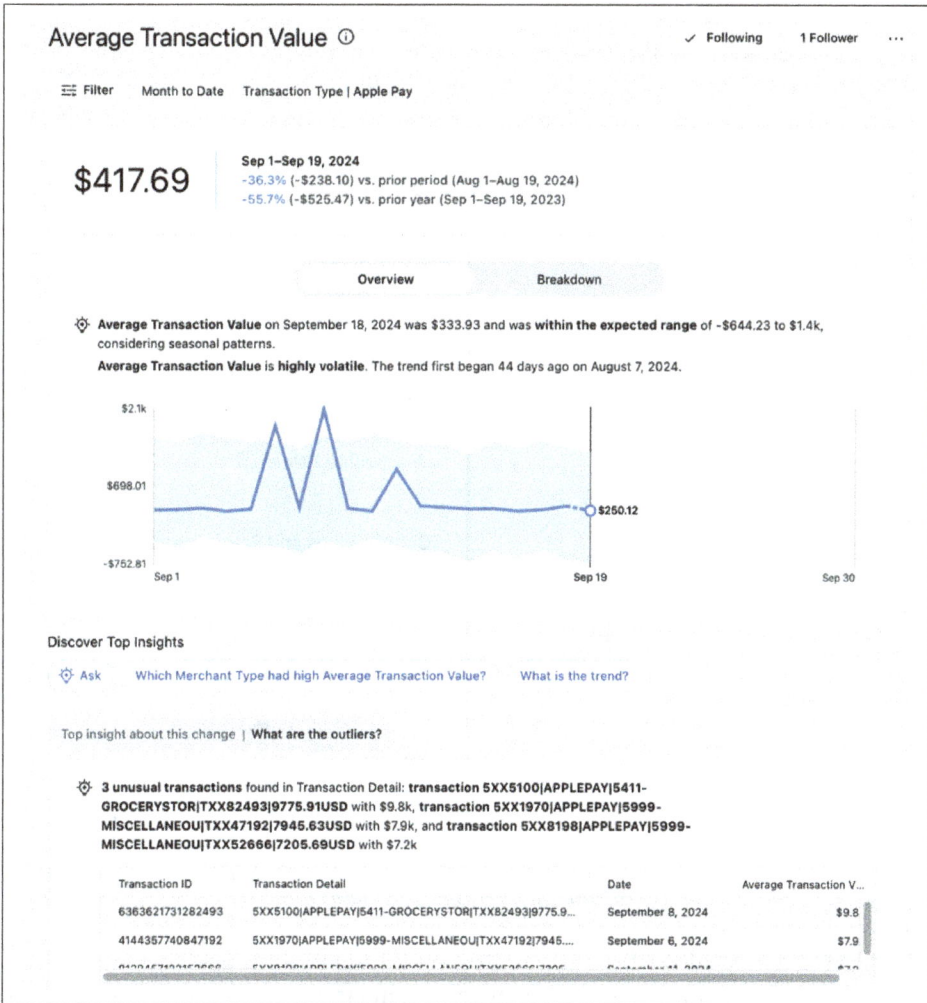

Figure 5-11. The full-size version of Average Transaction Value for Apple Pay (see a larger version of this figure online (https://oreil.ly/lait0511))

The top section of the metric is similar to what you've already seen, showing the time-bound trend and comparison you've come to know with Pulse. But the bottom section, where Discover Top Insights is located, reveals something new: outlier transactions that are well outside of the statistically normal values for Apple Pay transactions within the credit union for the month.

Traditionally, a significant amount of code and algorithm creation is necessary to achieve this type of anomaly or outlier detection within a data set, but Pulse makes it easy. Besides eliminating the need to define what an anomaly is, you can also customize the settings of the metric to ensure this is the only type of insight the product manager will be served up.

Figure 5-12 shows how the insights have been customized, relying heavily on the Record-Level Outliers feature. This includes defining a "Record identifier" and "Record identifier name," and filling in Singular and Plural nouns to describe the records. Additionally, because they're not relevant for the task at hand, all insights in the Contributions and Breakdowns section have been turned off.

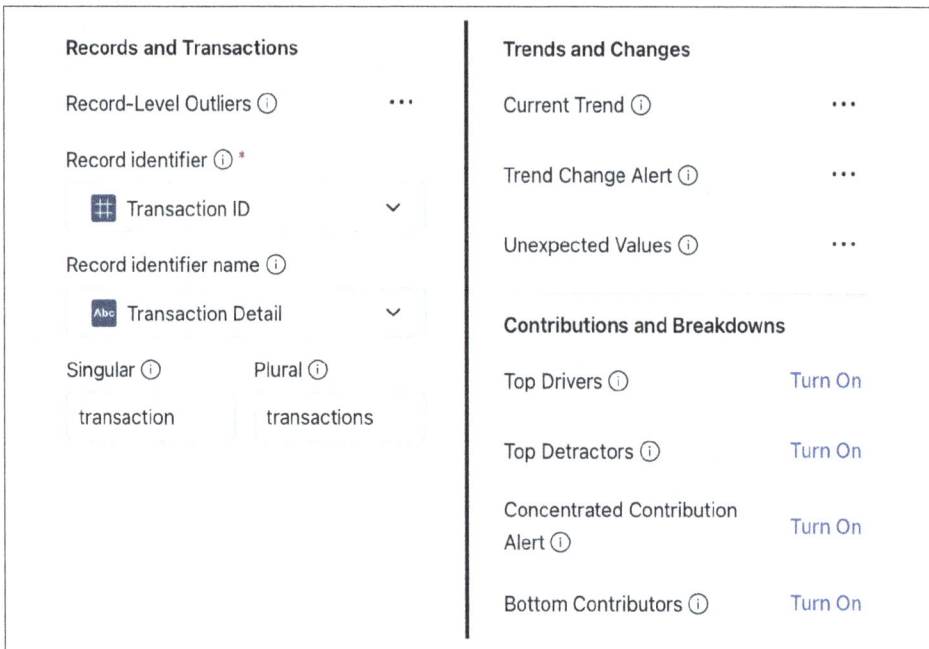

Figure 5-12. Insights configuration for Average Transaction Value

This level of tuning allows the outlier detection feature to shine and prominently displays the insights within the generated metric summaries. As you saw in Figure 5-10, the actual transactional data is displayed in a small data table, allowing immediate access to all the key information necessary to understand the anomalies. And although it is very unlikely that this anomaly detection will replace fraud detection applications used at the credit union, at least the product manager has real-time information available that can quickly shine a spotlight on anything atypical.

> Pulse allows you to specify the unique record identifier and an identifier name. Both pieces of information will show up in the outlier data table within the Discover Top Insights section. You can take advantage of this by constructing a concatenated field in your published data source, like Transaction Detail, containing all relevant information for the outliers.

Use Case: Healthcare

Up to now, you've seen how Tableau Pulse generates insights and can guide users on metrics they already follow. However, in this use case, you'll see how creating new metrics after an initial insight from Pulse can provide even more valuable tracking and alerting. You'll also see another example of how configuring specific insights for your audience helps support their needs.

This use case is for a pharmacy procurement manager at a large long-term care facility network. Since the organization is managing ongoing care for patients, monitoring fluctuating drug costs and supply is critical, especially with recent supply chain disruptions driving drug price instability. In the past, these costs have been reviewed on an annual basis, in alignment with supplier contract renewals, but the pharmacy procurement manager is looking for a more proactive way to assess risk, with the hope that the organization can have more negotiating power and alternative supplier options during contract renewal season.

Figure 5-13 shows a snippet of the data set, which includes the purchase history of various medications. Additional key fields are included, like the therapeutic category and usage information for the drug, the supplier name, and whether the supplier is a preferred vendor. These key fields will be included as adjustable filters in the metrics.

Date	Drug Name	Supplier	Supplier Type	Therapeutic Category	Therapeutic Use	Quantity Ordered	Total Value	Unit Cost
1/1/2023	Acetaminophen; Hydrocodone	Medline Pharmaceuticals	Preferred Supplier	Pain/Inflammation	Pain (Opioid analgesic)	61	19.07475	0.312700820
1/1/2023	Acetaminophen; Oxycodone	Ascend Medical Systems	Non-Preferred Supplier	Pain/Inflammation	Pain (Opioid analgesic)	38	10.622	0.279526316
1/1/2023	Albuterol	Unity Healthcare Supplies	Preferred Supplier	Respiratory	Asthma/Bronchospasm	77	20.39325	0.264847403
1/1/2023	Alendronate	Ascend Medical Systems	Non-Preferred Supplier	Metabolic/Endocrine	Osteoporosis	70	24.60125	0.351446429
1/1/2023	Allopurinol	Unity Healthcare Supplies	Preferred Supplier	Metabolic/Endocrine	Gout/Hyperuricemia	43	15.07125	0.350494186
1/1/2023	Alprazolam	Unity Healthcare Supplies	Preferred Supplier	Neurological/Psychiatric	Anxiety/Panic disorders	57	21.7705	0.381938596
1/1/2023	Amitriptyline	PrimaHealth Distributors	Non-Preferred Supplier	Neurological/Psychiatric	Depression/Neuropathic	43	14.75075	0.343040698
1/1/2023	Amlodipine	PrimaHealth Distributors	Non-Preferred Supplier	Cardiovascular	Hypertension/Angina	13	8.7525	0.673269231
1/1/2023	Apixaban	Horizon Pharma Group	Preferred Supplier	Cardiovascular	Blood thinner (Anticoag)	67	49.60775	0.740414179
1/1/2023	Aripiprazole	Vitality Medical Group	Non-Preferred Supplier	Neurological/Psychiatric	Schizophrenia/Bipolar D	59	17.48425	0.296343220

Figure 5-13. A subset of drug-purchasing history

To help identify price fluctuation risks or changes from available suppliers, the manager already follows three metrics, shown in Figure 5-14: Total Cost, Units Ordered, and Unit Cost. These will serve as the launching point for further investigation.

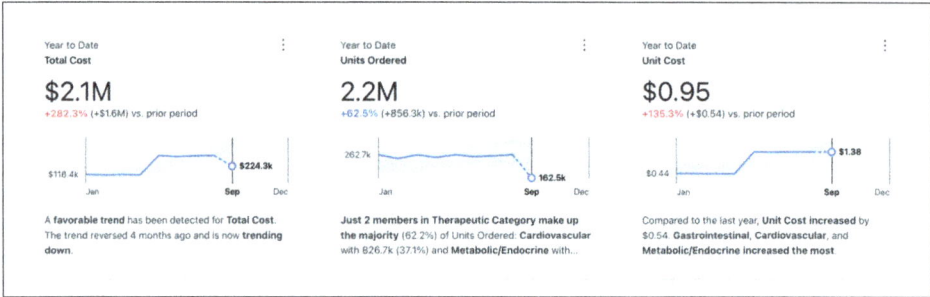

Figure 5-14. Pulse metrics that the pharmacy procurement manager follows (see a larger version of this figure online (https://oreil.ly/lait0514))

Focusing on the Units Ordered metric, its insights have been limited to those that can help identify the largest areas of risk and spot a situation where a supplier may have too much influence over the price of drugs ordered. Figure 5-15 shows the customizations. The Total Cost and Unit Cost metrics have been configured similarly, with one additional insight turned on: Current Trend.

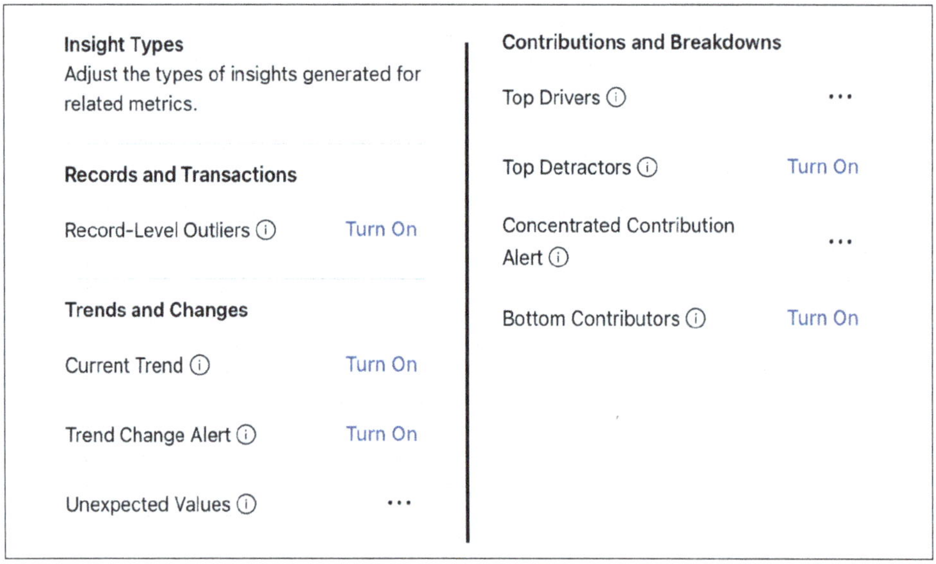

Figure 5-15. The insight configurations for Units Ordered

With the insights specifically configured, it becomes clear where the manager should focus her immediate efforts: on cardiovascular drugs, a therapeutic category mentioned in each of the three metric summaries. To see if this is really an issue, she can use the adjustable filters to set each metric to include only cardiovascular medications. Figure 5-16 shows what surfaces after she limits the Units Ordered metric, unearthing that the company is ordering more cardiovascular drugs from nonpreferred suppliers. Furthermore, 54.1% of these medications are procured from just three suppliers, representing a potential supplier concentration risk in their supply chain.

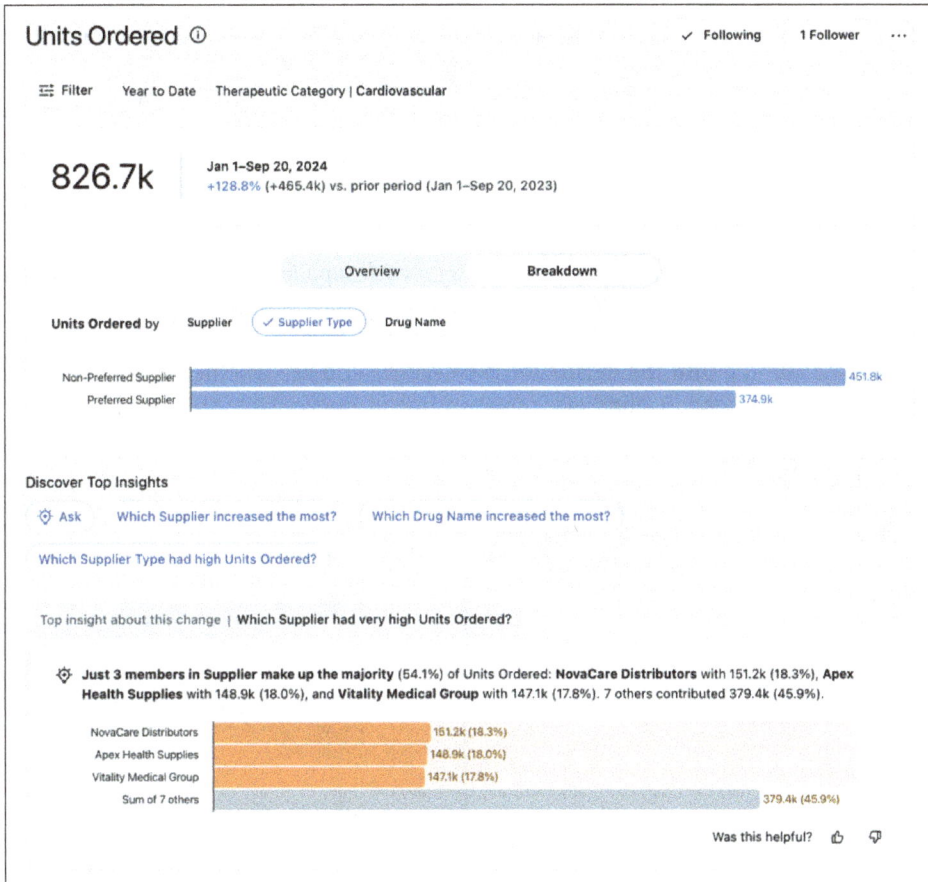

Figure 5-16. Cardiovascular Units Ordered metric breakdown (see a larger version of this figure online (https://oreil.ly/lait0516))

Armed with this new information about the suppliers, she can now build even more pointed metrics to see how Unit Cost has been affected by this finding. Figure 5-17 shows the metrics she's created, allowing her to track the unit cost of cardiovascular drugs for both preferred and nonpreferred suppliers separately.

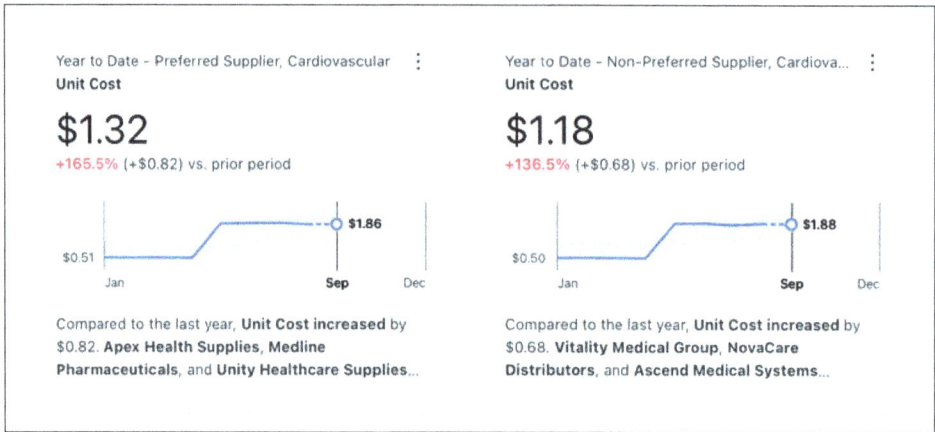

Figure 5-17. Cardiovascular Unit Cost metrics for each supplier type (see a larger version of this figure online (https://oreil.ly/lait0517))

Finally, she can dig into the details for each metric and see whether this increase in cost is uniform across all suppliers, representing an unavoidable shift in the market, or it's a supplier-related issue that can be mitigated. Figure 5-18 shows the breakdown for each metric.

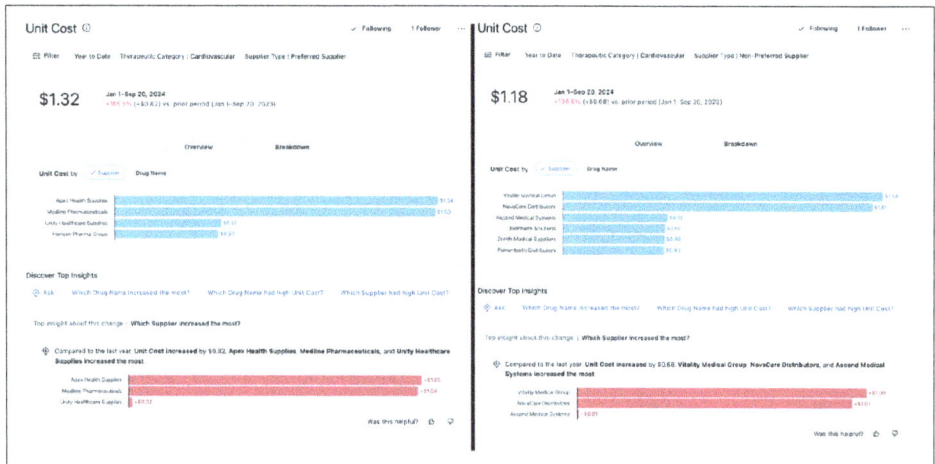

Figure 5-18. Breakdown for each supplier type Unit Cost metric (see a larger version of this figure online (https://oreil.ly/lait0518))

With just a few clicks, she's drilled down and found actionable insights. The rising cost in cardiovascular drugs appears to be limited to a handful of suppliers, both preferred and nonpreferred. Knowing this, she can not only track the pricing fluctuations for these specific suppliers, but also feed this information back to the company's medication buyers to immediately change their purchasing strategies. She has also followed all these metrics, ensuring they're closely tracked until the issue is resolved.

Use Case: Pulse Utilization

To wrap up this chapter, you're going to see how Tableau Pulse can be used to monitor itself. Recall from Chapter 3 that multiple data sets are included in the Admin Insights project that can help you understand how end users are working with Pulse. You can refer to that chapter to refresh your memory on the fields available and their definitions.

Site Content Data Source

Using the Site Content data source available in Admin Insights, Figure 5-19 shows five metric definitions you can create to get started. As you begin implementing Pulse in your Tableau environment, these will help you understand user adoption. They will also give you insight into the maturity of your Pulse environment by identifying how many data sources are in use and an understanding of how many child metrics are created from definitions. Instructions on how to construct these metrics are described in Table 5-1.

> Since these metrics are based on a creation date, you may want to specify larger periods of time for aggregation.

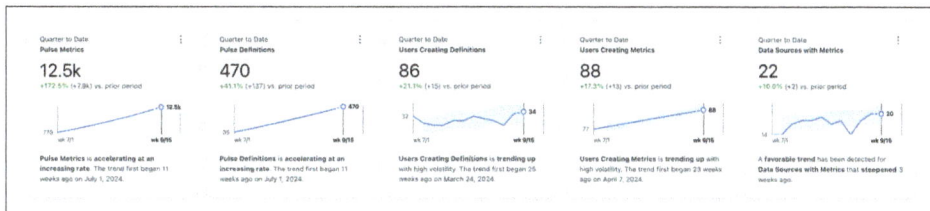

Figure 5-19. Metrics showing Pulse utilization (see a larger version of this figure online (https://oreil.ly/lait0519))

Table 5-1. Metric definitions for monitoring Pulse adoption

Name and description	Configurations
Pulse Metrics Measures the number of unique Pulse metrics created within a given time period	Measure and Aggregation: *Count of Item LUID* Definition Filters: *Item Type = Metric* Time Dimension: *Created At (Local)* Adjustable Filters: *Owner Email*
Pulse Definitions Measures the number of unique Pulse metric definitions created within a given time period	Measure and Aggregation: *Count of Item LUID* Definition Filters: *Item Type = Metric Definition* Time Dimension: *Created At (Local)* Adjustable Filters: *Owner Email*
Users Creating Definitions Measures the number of unique users creating metric definitions within a given time period	Measure and Aggregation: *Distinct Count of Owner Email* Definition Filters: *Item Type = Metric Definition* Time Dimension: *Created At (Local)* Adjustable Filters: *Owner Email*
Users Creating Metrics Measures the number of unique users creating metrics within a given time period	Measure and Aggregation: *Distinct Count of Owner Email* Definition Filters: *Item Type = Metric* Time Dimension: *Created At (Local)* Adjustable Filters: *Owner Email*
Data Sources with Metrics Measures the number of unique data sources that have metrics within a given time period	Measure and Aggregation: *Distinct Count of Metric Definition Data Source ID* Time Dimension: *Created At (Local)* Adjustable Filters: *Owner Email*

I recommend turning off the Record-Level Outliers, Top Detractors, and Bottom Contributors insights, as they don't add much value to understanding how Pulse is growing within your Tableau environment.

> For richer analytics on Pulse utilization, you can combine other data sources available in Admin Insights by using Prep Builder. For example, you can join the Site Content data source with the TS Users data source to include the user's site role and license type. Similarly, you can include the Groups a user is a member of by joining to the Groups data source.

Subscriptions Data Source

You can use the Subscriptions data source to understand the following activity associated with users and metrics. Figure 5-20 shows three metrics you can use to understand user activity. This type of activity tracking can be useful in understanding whether any metrics suddenly have a high level of interest. Table 5-2 provides instructions on how to construct these metrics.

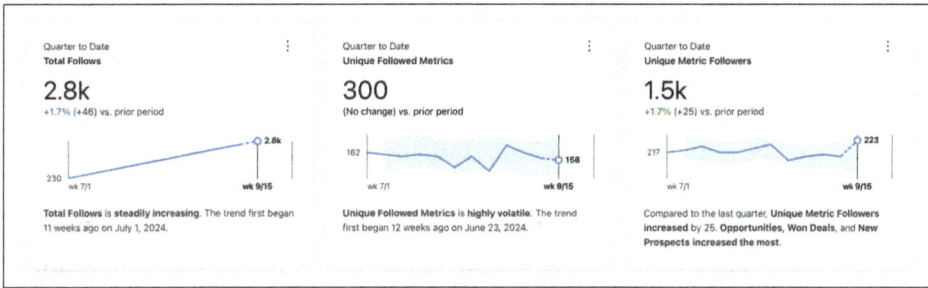

Figure 5-20. Metrics showing Pulse activity (see a larger version of this figure online (https://oreil.ly/lait0520))

Table 5-2. Metric definitions for measuring Pulse following activity

Name and description	Configurations
Total Follows Measures the total number of follows for Pulse metrics created within a given time period	Measure and Aggregation: *Count of Subscriber Email* Definition Filters: *Item Type = Metric* Time Dimension: *Created At (Local)* Adjustable Filters: *Item Name*
Unique Followed Metrics Measures the number of unique Pulse metrics followed within a given time period	Measure and Aggregation: *Distinct Count of Item LUID* Definition Filters: *Item Type = Metric* Time Dimension: *Created At (Local)* Adjustable Filters: *Subscriber Email*
Unique Metric Followers Measures the number of unique users following Pulse metrics within a given time period	Measure and Aggregation: *Distinct Count of Subscriber Email* Definition Filters: *Item Type = Metric* Time Dimension: *Created At (Local)* Adjustable Filters: *Item Name*

Similar to the metrics created from Site Content, I recommend turning off the Record-Level Outliers, Top Detractors, and Bottom Contributors insights.

> Since the time dimension used from Subscriptions is a creation date, you'll be seeing point-in-time activity and not actual values. As an example, the Total Follows metric described in Table 5-2 provides the total follows within the metric time period, but not the total follows active in the environment.

Summary

You've now seen four distinct use cases for Tableau Pulse. In the sales example, you saw how an existing dashboard can be reimagined as Pulse metrics. As the example was explored, you saw how Pulse metrics provide a very direct insights experience compared to the self-service and exploration afforded in the dashboard. In both the finance and healthcare examples, you saw how fine-tuning the types of insights generated can be used for anomaly detection and concentrated risk detection, respectively. Additionally, within the healthcare example, you journeyed along with the pharmaceutical procurement manager through the creation of new metrics, ultimately leading to actionable intelligence for buyers. Finally, you saw a variety of metrics that can be created from Admin Insights that can immediately help you understand both user adoption and activity as your organization uses Pulse.

Chapter 6 shifts gears and explores Pulse from the angle of a developer. You'll learn how to integrate Pulse with Slack, embed Pulse metrics within Salesforce or a custom web application, and understand key details about the APIs available to support your development work.

Integrating and Extending Tableau Pulse

In the previous chapters, you saw how Tableau Pulse works for metric authors and metric end users; now it's time to learn about how to integrate Tableau Pulse with Slack, Salesforce customer relationship management (CRM), and custom web applications or web pages. In this chapter, you'll also learn about developer tools, like the Tableau REST API, allowing you to programmatically create and administer metric definitions, metrics, insights, and digest subscriptions.

> This chapter contains advanced developer concepts that may be outside of the scope of a data analyst's or Tableau developer's day-to-day role. The purpose of this chapter is to expose you to these concepts, both demonstrating and explaining the advanced features available when using Tableau Pulse. It is advisable that you partner with a developer who specializes in Salesforce, web development, and/or working with APIs when implementing.

Slack Integration

In Chapter 4, you learned how an end user can enable Slack to start receiving Pulse digests for the metrics they follow. However, before this feature is available to them, administrators for both Tableau Cloud and your Slack workspace will need to enable connectivity between the two platforms.

> For organizations using Microsoft Teams, Tableau has released an app very similar to the Slack app. More information about this, including how to enable it within your organization, can be found at the Tableau App for Microsoft Teams (*https://oreil.ly/61Qqc*) launch blog.

Tableau Cloud Administration

To begin, you'll want to start in your Tableau Cloud environment. Remember, the user who is making these modifications to the Tableau Cloud site will need to have a license role of Site Administrator Creator or Site Administrator Explorer. Once you've determined that you have the right role, navigate over to the Settings section. From there, access the Integrations section and scroll down to Slack Connectivity as highlighted in Figure 6-1.

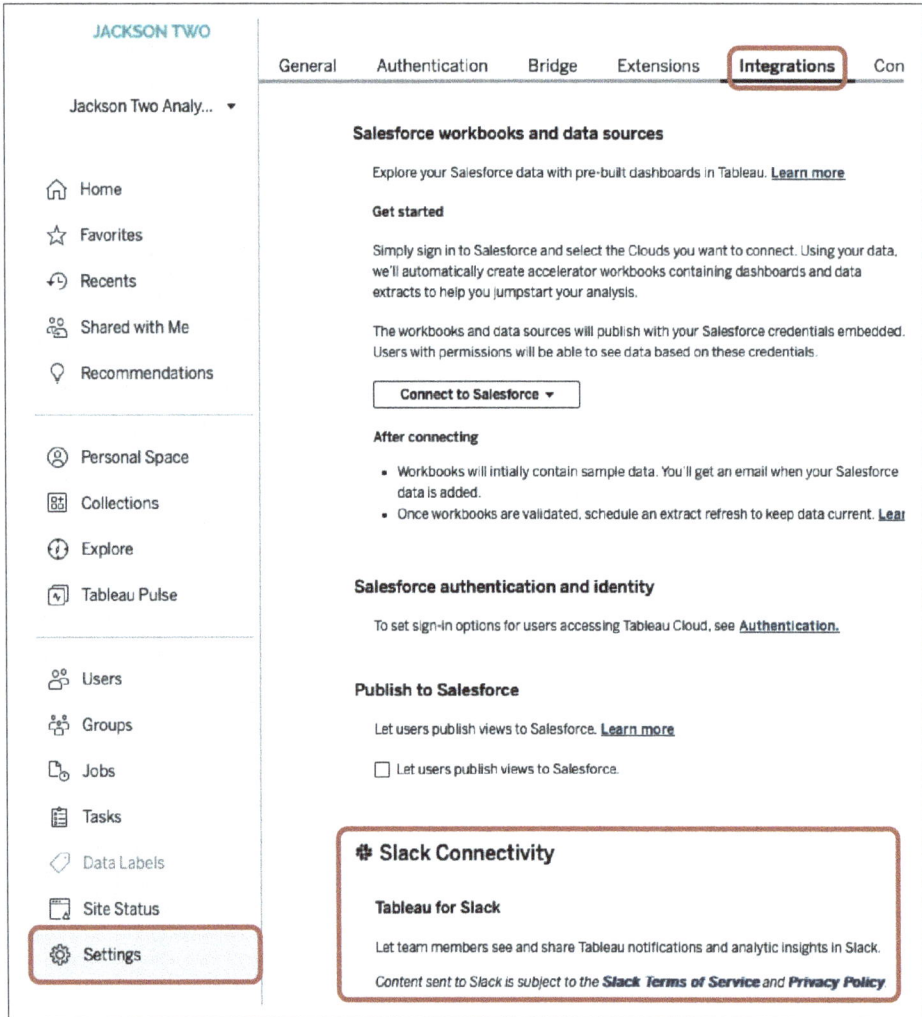

Figure 6-1. The Integrations area of Settings in a Tableau Cloud site (see a larger version of this figure online (https://oreil.ly/lait0601))

You and your end users can do more with Tableau and Slack than just receive Pulse digests. Enabling integration with Slack also allows users to do the following:

- Receive notifications when content is shared with them
- Receive notifications when they are mentioned in a comment
- Receive notifications when a data-driven alert is triggered
- See a preview of the visualization when the URL is shared in channels or DMs
- Search for workbooks and views in channels or DMs
- Access Tableau Recents and Favorites by using the Tableau App

Enabling integration with Slack requires agreeing to allow an intermediary called the *Salesforce Slack Integration Proxy* to route data between Tableau and Slack, as shown in Figure 6-2.

Figure 6-2. Checkbox to enable Salesforce Slack Integration Proxy

After the Salesforce Slack Integration Proxy has been enabled, you'll be able to click Connect to Slack. A new web page will pop up with the request from Tableau to gain permission to the Slack workspace. This pop-up includes details on what content Tableau will be able to view from Slack, as well as the types of actions it will be allowed to perform in Slack on a user's behalf. Figure 6-3 shows the detailed permissions pop-up for a Slack workspace where Slack administrators must first approve an app prior to installation.

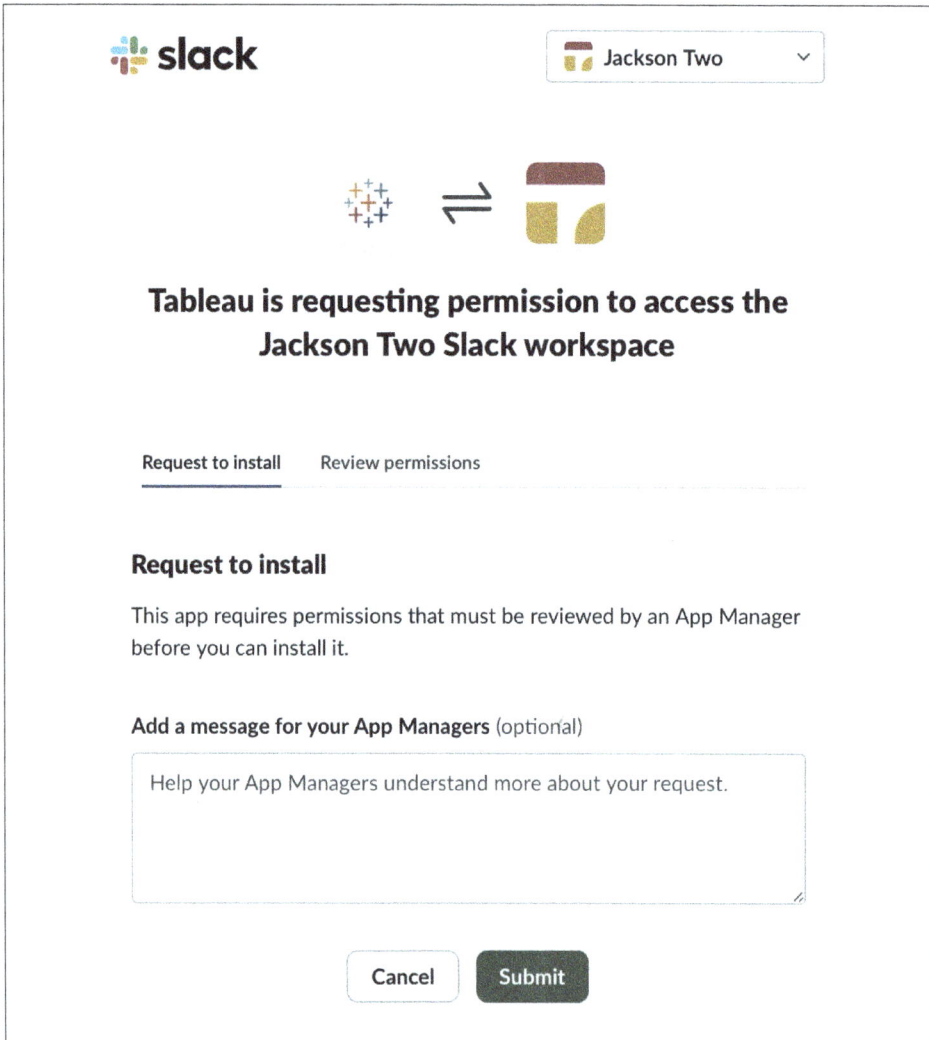

Figure 6-3. Pop-up window request for Tableau Cloud to access a Slack workspace

Clicking Submit causes the window to disappear, and there will be an approval request to submit to your Slack administrator. A Slack administrator will then have the option to approve the app for the entire workspace. Once it has been approved, head back to Tableau Cloud and repeat the Connect to Slack process again. This time, you'll see an approval message instead of a request message, as shown in Figure 6-4. Click Allow to enable the connection.

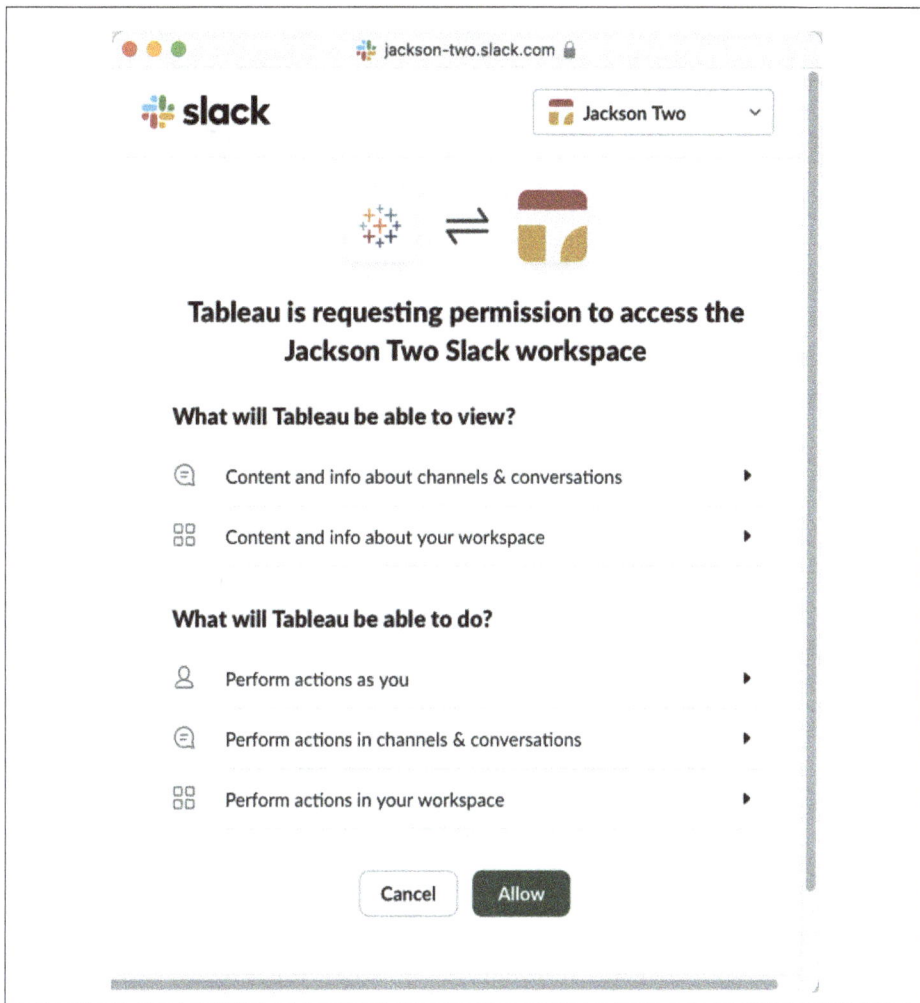

Figure 6-4. Request to connect Slack to Tableau Cloud

When you head back to the Integrations section of Settings in Tableau Cloud, you will now see your Slack workspace connected, as shown in Figure 6-5.

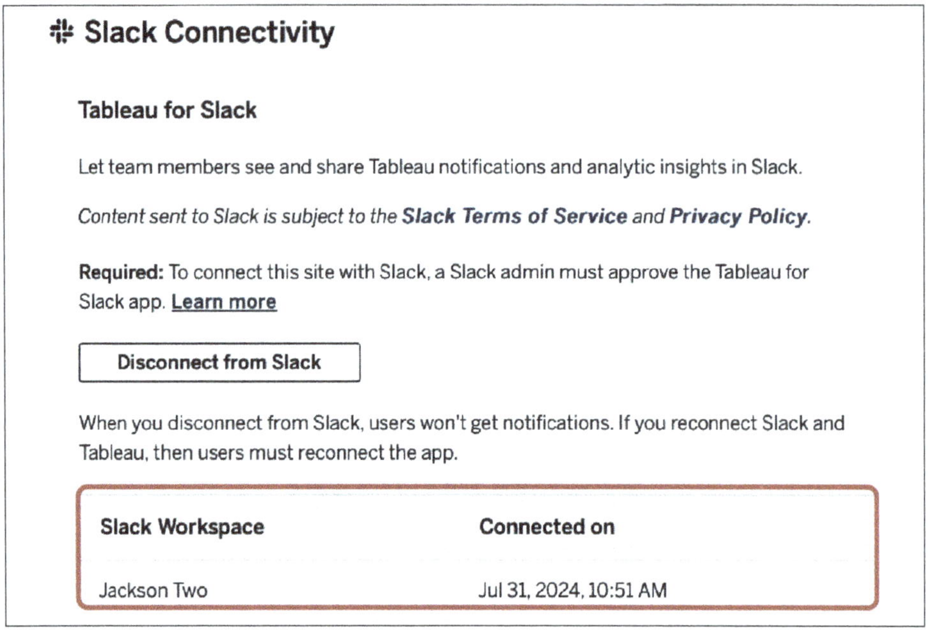

⌗ Slack Connectivity

Tableau for Slack

Let team members see and share Tableau notifications and analytic insights in Slack.

*Content sent to Slack is subject to the **Slack Terms of Service** and **Privacy Policy**.*

Required: To connect this site with Slack, a Slack admin must approve the Tableau for Slack app. **Learn more**

> Disconnect from Slack

When you disconnect from Slack, users won't get notifications. If you reconnect Slack and Tableau, then users must reconnect the app.

Slack Workspace	Connected on
Jackson Two	Jul 31, 2024, 10:51 AM

Figure 6-5. The Integrations section with the connected Slack workspace

If you happen to also be a Slack workspace administrator with the ability to add Slack apps, you'll notice that the Tableau app is automatically installed in your Slack workspace after you click Connect to Slack the first time.

Slack Administration

If you're a Slack workspace administrator where app management is restricted, you can enable the Tableau app for your workspace with a few simple clicks. To proactively allow the Tableau app, head to the *Slack App Directory*, which can be accessed by selecting "Add apps" from the sidebar in your Slack workspace. Once you're in the directory, search for the Tableau app. Figure 6-6 shows what the Tableau app for Slack information page looks like.

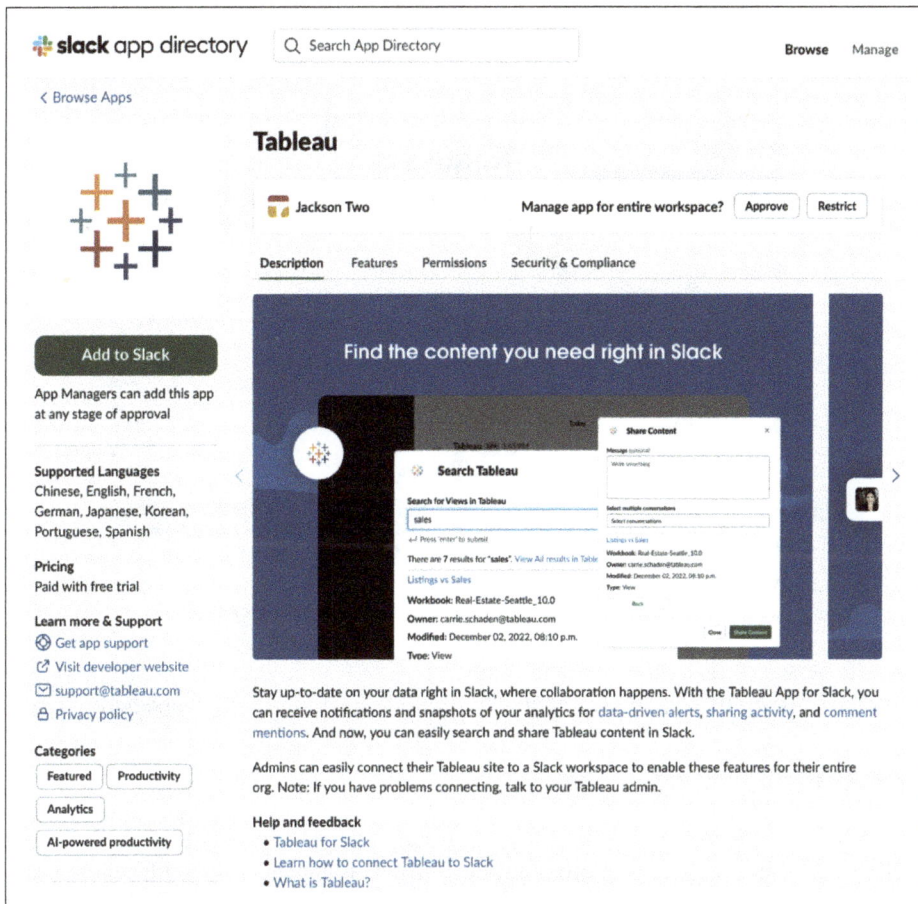

Figure 6-6. The Tableau app for Slack information and management page

From here, you can click the Approve button. This will allow Tableau Cloud administrators to complete the connection process previously mentioned. Additionally, it will allow end users to add the Tableau app to their Slack workspace, as described in Chapter 4.

Alternatively, you can navigate to the Manage section of the Slack App Directory to review requests and approve using the action from the triple dot actions menu, as shown in Figure 6-7, or respond to the request directly inside Slack via a DM from Slackbot.

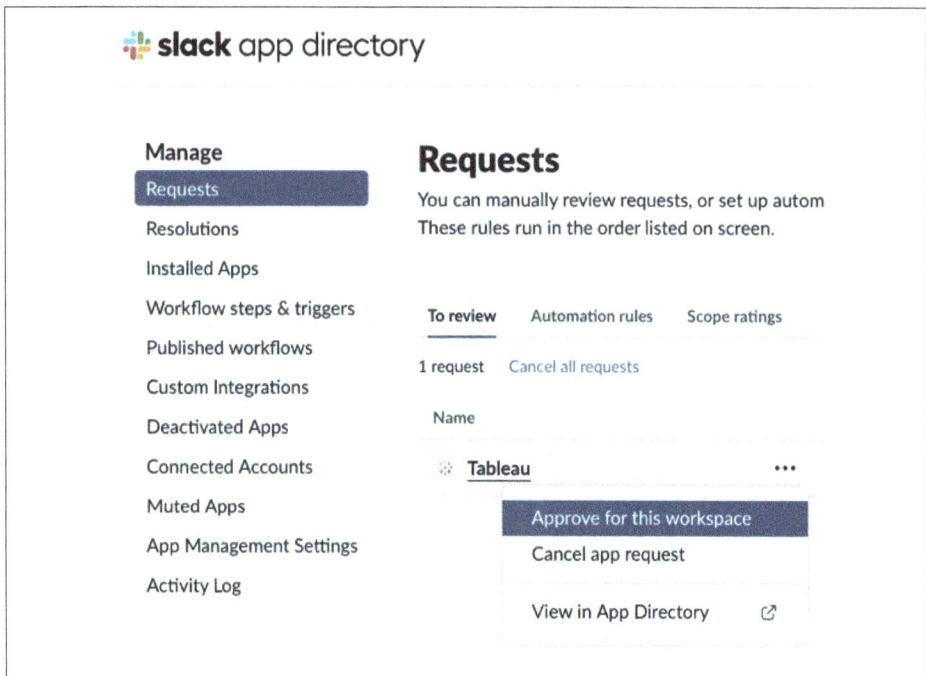

Figure 6-7. Alternative method to approve a Tableau app in the Slack workspace

Embedding Pulse

One of the unique features of the Tableau product line is the ability to embed visualizations and Pulse metrics into other applications. This capability extends where users can access, view, and interact with Tableau content, specifically outside the Tableau Cloud environment. This feature allows users to take full advantage of Tableau without leaving the application they work in—saving time and giving them direct access to meaningful data, all within the flow of work.

Salesforce CRM

The core software in the Salesforce ecosystem is its CRM. This web-based application is where organizations can store account and contact information, track leads, and monitor completed sales. For many organizations, members of their sales teams spend most of their daily working time within the CRM.

Since members are spending so much time within the CRM, it makes sense to bring Pulse metrics and insights into the platform. This gives CRM users the potential to monitor analytics and access metrics that may not be directly stored or visible in the CRM. To get started embedding Tableau Pulse inside the CRM, you'll need administrative permissions for both Tableau Cloud and Salesforce CRM.

Setup in Salesforce

The best place to start enabling the embedding process for Tableau Pulse into Salesforce CRM is within Salesforce itself. There are two configuration screens that a Salesforce administrator will navigate through to enable the integration between the two applications. First is setting up Tableau Cloud and Tableau Pulse as a trusted URL. To do this, click the settings (cog) icon in the upper-right corner of Salesforce and select Setup. This will take you to the Setup home page where you can then access different sections and begin configuration.

Search for Trusted URLs in the Quick Find box on the left. Figure 6-8 shows the Trusted URLs page.

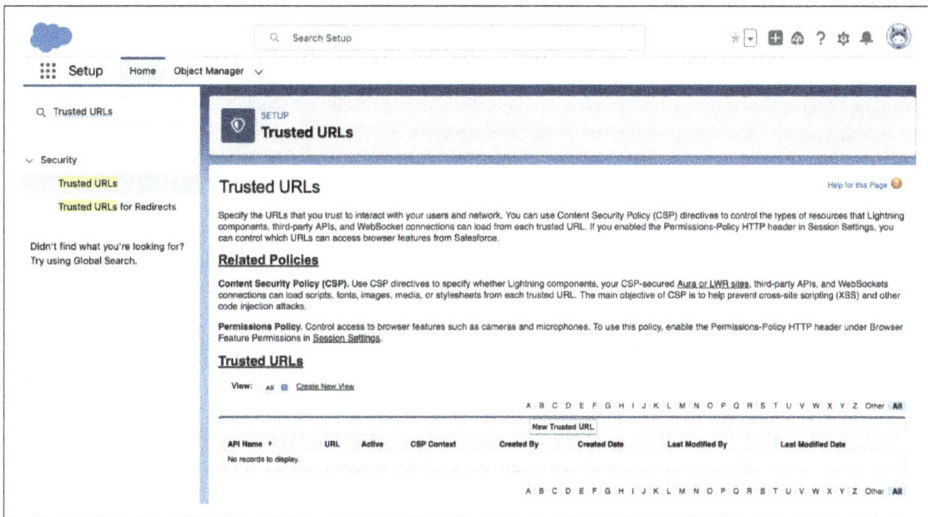

Figure 6-8. The Trusted URLs setup page in Salesforce (see a larger version of this figure online (https://oreil.ly/lait0608))

From here, create a new Trusted URL entry for Tableau Cloud. Click New Trusted URL (bottom of Figure 6-8) to begin entering in a new URL entry, as shown in Figure 6-9. Fill in the following information:

API Name
> Choose a friendly name to describe the URL, which cannot include spaces.

URL
> The fully qualified hostname of the application. For Tableau Cloud and Pulse, this is the root URL for the application.

Description
> Give a brief summary of the application represented by the URL.

Enable the following Content Security Policy (CSP) settings:

- Set CSP context to All.
- Ensure that all checkboxes within the CSP Directives settings are checked.

The last section of the entry page includes enabling camera and microphone access, which isn't necessary (or used) for Tableau integrations and can remain unchecked.

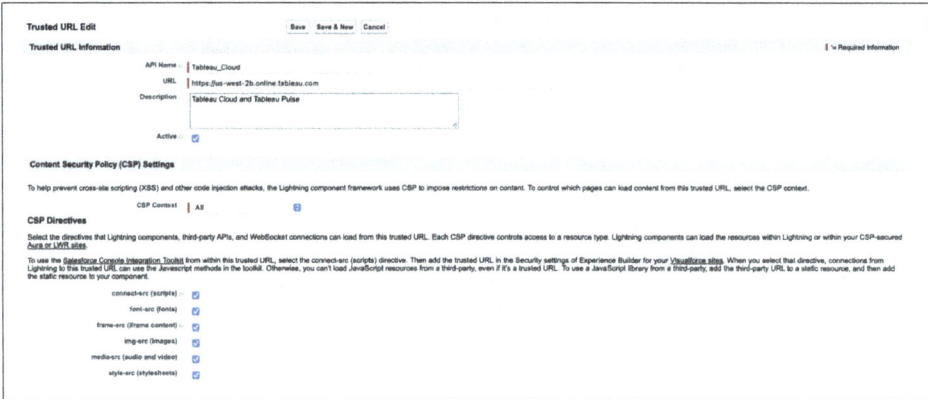

Figure 6-9. A completed Trusted URL entry in Salesforce (see a larger version of this figure online (https://oreil.ly/lait0609))

Next, enable and configure the Tableau Lightning Web Components (LWCs) within Salesforce by searching for Tableau Embedding in the Quick Find section within Setup. If it's the first time you've navigated to this screen, there may be an option to check a box next to Turn on Tableau View Lightning Web Component.

Complete the following configurations, as shown in Figure 6-10:

- Click the checkbox to choose "Turn on token-based single sign-on authentication," which allows users to be authenticated between the two applications.
- Specify the "Select Tableau User Identity field" to be used for the SSO process. This is a field within the Salesforce application that aligns with a user's username in Tableau Cloud.
- Copy the Issuer URL and JWKS URI. These two pieces of information will be used later when configuring Tableau Cloud.
- Create a new Tableau Host Mapping entry. This will include the site URL and site ID, described in the next section. Set the Tableau Host Type to Tableau Cloud.

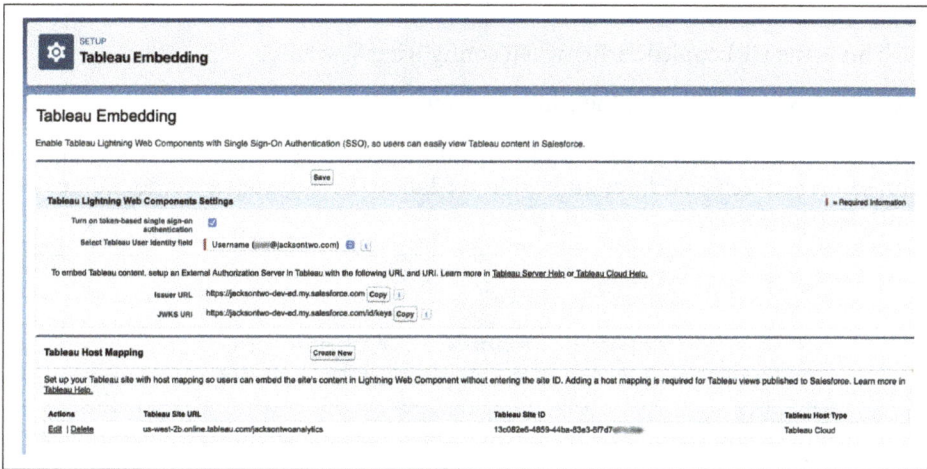

Figure 6-10. Completed Tableau Embedding configuration settings in Salesforce (see a larger version of this figure online (https://oreil.ly/lait0610))

Setup in Tableau Cloud

Once you've initiated the configurations within Salesforce CRM, you'll need to set up Salesforce as a connected app in Tableau Cloud. To do this, go to Settings and select Connected Apps. Click the New Connected App dropdown and select OAuth 2.0 Trust, as shown in Figure 6-11.

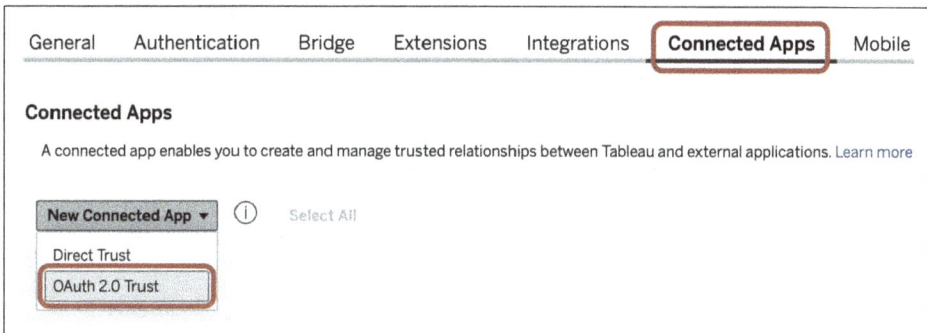

Figure 6-11. Specifying the connected app type from the Settings menu in Tableau Cloud

You'll be prompted to enter the following information:

Connected App Name
    This is a friendly name to denote which application the connection is for.

Issuer URL
    This is the issuer URL that was copied earlier when configuring Salesforce.

JWKS URI

This is the URI copied earlier when configuring Salesforce.

Click the checkbox to enable the connected app.

Click the Create button to return to the Connected Apps section and complete the connection between Salesforce and Tableau Cloud. The completed configurations are shown in Figure 6-12.

Create Connected App

Establish a trusted relationship with an external authorization server. Enter information about the external authorization server that will generate the JSON web token (JWT).
Learn more

Connected app name	Salesforce
Issuer URL	https://jacksontwo-dev-ed.my.salesforce.com
JWKS URI	https://jacksontwo-dev-ed.my.salesforce.com/i

☑ Enable connected app

[Cancel] [**Create**]

Figure 6-12. Completed configurations for the connection with Salesforce

After the entry is created, you'll be presented with the details along with the Site ID, as shown in Figure 6-13. It includes an option to copy the Site ID, which is necessary for Tableau host mapping in Salesforce. Pass this information back to your Salesforce administrator or, if you're also a Salesforce administrator, update it yourself.

> The Site ID is the unique identifier for your Tableau Cloud environment. This information is also used when adding Tableau components to Salesforce, so it's a good idea to keep this information handy.

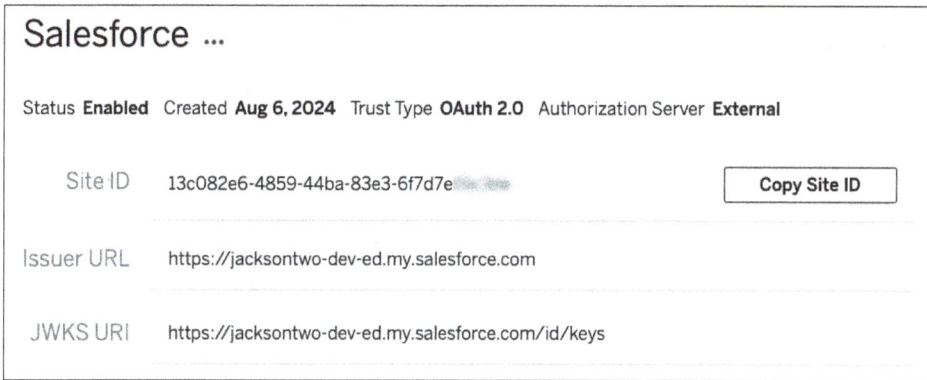

Salesforce ...

Status **Enabled** Created **Aug 6, 2024** Trust Type **OAuth 2.0** Authorization Server **External**

Site ID	13c082e6-4859-44ba-83e3-6f7d7e⬛ ⬛
Issuer URL	https://jacksontwo-dev-ed.my.salesforce.com
JWKS URI	https://jacksontwo-dev-ed.my.salesforce.com/id/keys

Figure 6-13. Connected app details in Tableau Cloud with site ID

You can also use the URL from this web page to copy the Tableau site URL. Copy the URL up to the forward slash after the site name, as highlighted in Figure 6-14.

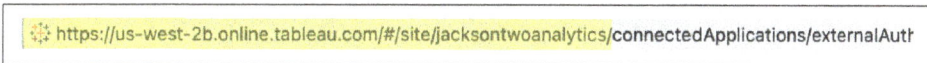

https://us-west-2b.online.tableau.com/#/site/jacksontwoanalytics/connectedApplications/externalAuth

Figure 6-14. The Tableau site URL for host mapping

This process establishes the relationship between the two applications by using trusted tokens, which is necessary to embed and view Tableau Pulse pages in Salesforce. The trusted tokens will be used to skip Tableau Cloud user authentication while in Salesforce. The last step in the process of embedding Tableau Pulse is building a new page or incorporating the Tableau Pulse component into an existing Salesforce page.

Standalone embedded Pulse page in Salesforce

To begin this process, head back to Salesforce. While in Setup, search for Lightning App Builder by using the Quick Find bar. Once there, click New to begin building a new page. Inside the builder, select App Page to start the process of creating a new tabbed page in Salesforce. Give the page a label, such as Tableau Pulse. Finally, specify the layout you want to use. Since this is a standalone page, you can select one region.

Now from the lefthand side, search or select Tableau Pulse from the available components list. Once you've located it, drag and drop it to the top section of the page area in the builder, as shown in Figure 6-15.

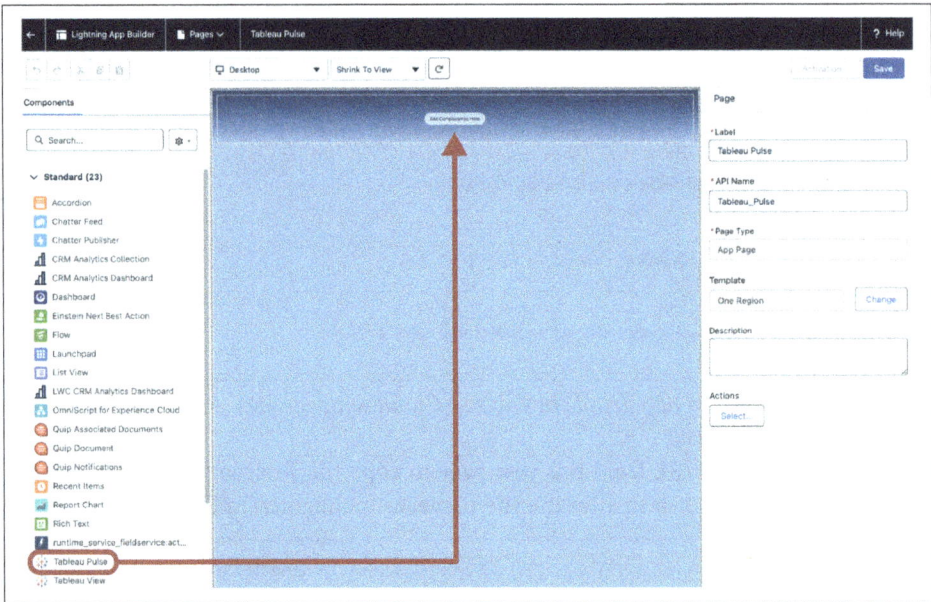

Figure 6-15. Drag and drop the Tableau Pulse component to the top section (see a larger version of this figure online (https://oreil.ly/lait0615))

Once the Pulse component has been dragged onto the page, specify the Page, Site ID, and Height of the component. Figure 6-16 shows a completed configuration set to the Tableau Pulse home page.

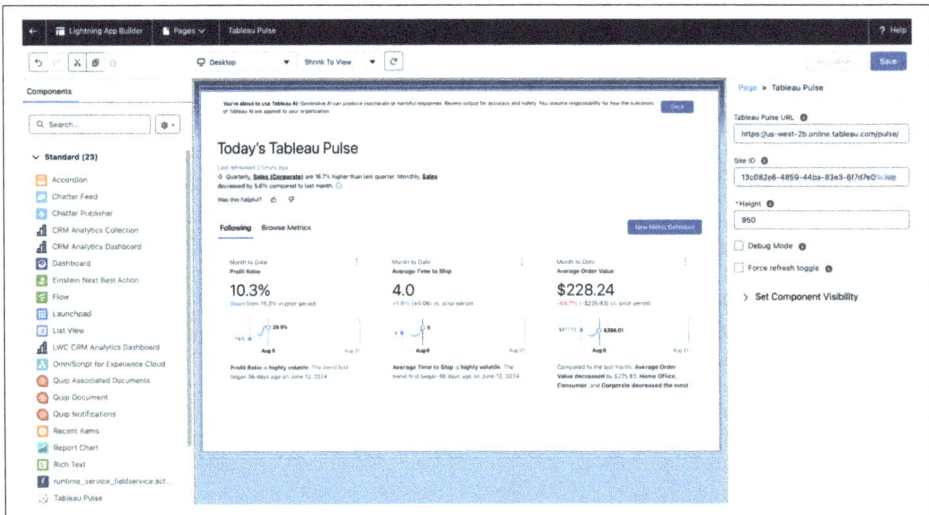

Figure 6-16. Completed configuration of the Tableau Pulse component on a Lightning page (see a larger version of this figure online (https://oreil.ly/lait0616))

Click Save, and then, when prompted, click Activate to activate the page. It will also ask you to specify where you want the page to be visible in the Lightning Experience and Mobile Navigation. Figure 6-17 shows the page saved as a tab within the Sales app.

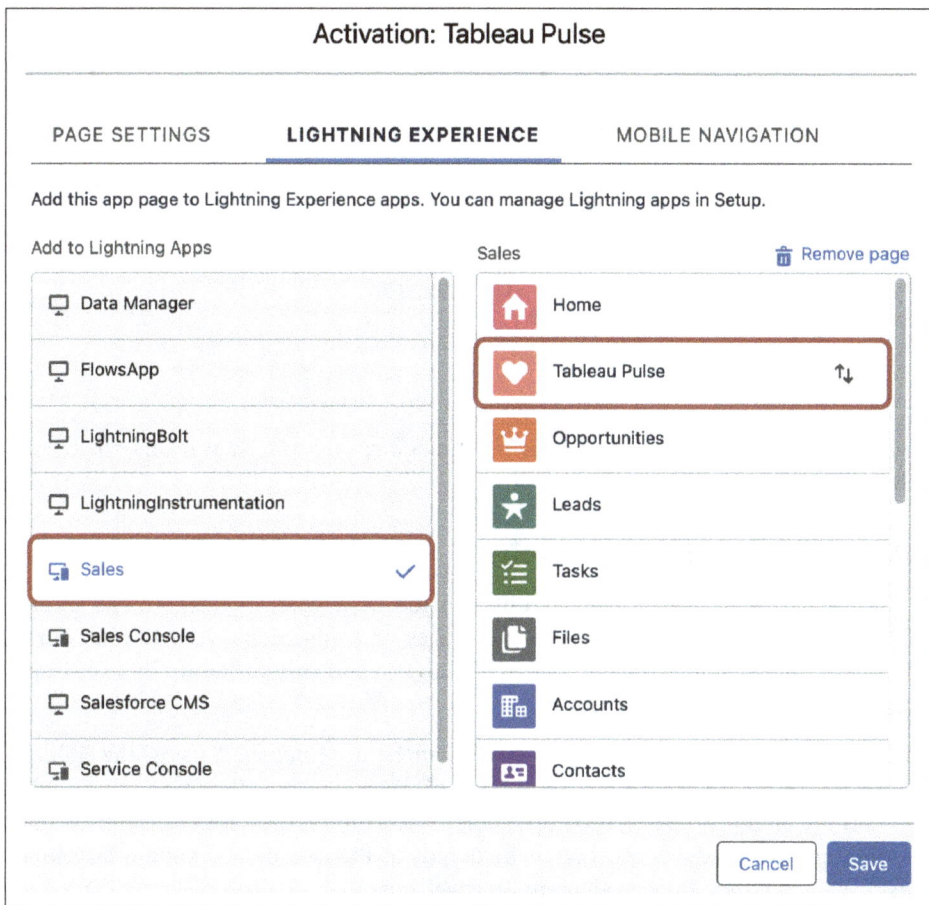

Figure 6-17. Inclusion of the Tableau Pulse page within the Sales app

Once the page has been added to one or more apps, Salesforce users will be able to access their Pulse summary and create and follow new metrics without leaving Salesforce, as shown in Figure 6-18.

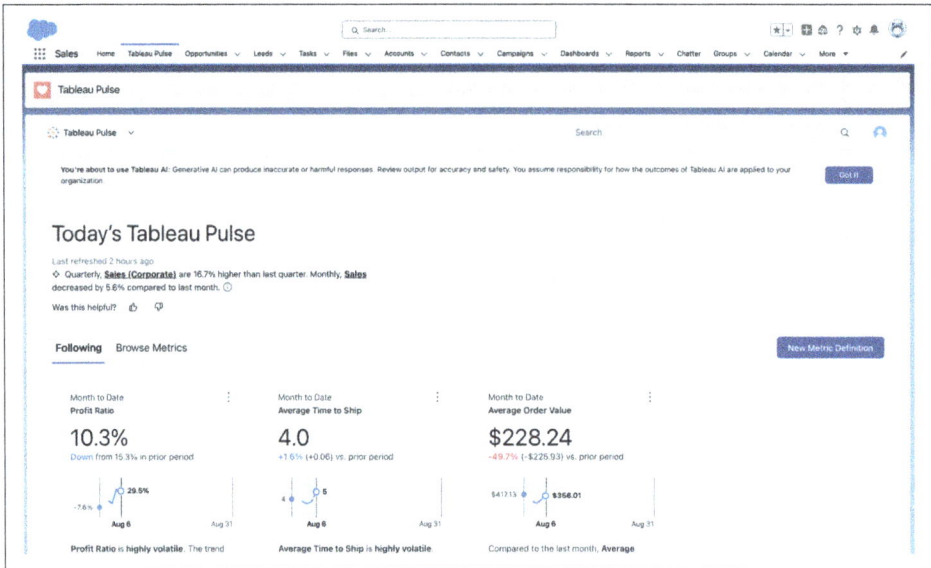

Figure 6-18. Tableau Pulse embedded within Salesforce (see a larger version of this figure online (https://oreil.ly/lait0618))

Custom Web Pages and Applications

In addition to embedding Tableau Pulse inside Salesforce, developers can embed Tableau Pulse metrics inside custom web pages or applications. As with Salesforce, this allows developers to include the rich analytics and insights from Tableau Pulse, where end users are already working.

Embedding Tableau Pulse into web pages can be accomplished using the Tableau Embedding API v3. This is the latest version of developer tools for embedding, which includes the ability to use web components. Web components are customizable and reusable encapsulated HTML tags, which reduce the amount of code a developer must use to include a Tableau element on a web page.

To begin, you must first link to the Embedding API v3 library on your web page. Tableau recommends matching the link location of the library to the Tableau version in use at your organization. Since Tableau Pulse is available only on Tableau Cloud, a best practice would be to link to the library available for Tableau Cloud. Tableau has several instances of Tableau Cloud running, which are location (geography) dependent. These instances, called *pods*, can be determined by looking at the subdomain of your Tableau Cloud's URL. Example 6-1 shows the code block link, where `Tableau-Pod` should be replaced with the pod name your Tableau Cloud instance is running on.

Example 6-1. Link to the Tableau Cloud embedding library

```
<script type="module" src="https://Tableau-Pod.online.tableau.com/
javascripts/api/tableau.embedding.3.latest.min.js"></script>
```

Next, you can use the Tableau Pulse web component to define and include the Tableau Pulse metric. This includes five properties for custom configuration. Table 6-1 lists the properties and how they impact the display of the Tableau Pulse web component.

Table 6-1. Properties for Tableau Pulse web component

Property	Description	Accepted values	Required
height	The height of the component on the web page; if not specified, defaults to the published height of the metric	Any valid CSS size specifier. For example: 800 px	No
width	The width of the component on the web page; if not specified, defaults to the published width of the metric	Any valid CSS size specifier. For example: 600 px	No
disable-explore-filter	Indicates whether the Explore Filters button is hidden; defaults to false, which includes the Explore Filters button	A Boolean value of true or false.	No
layout	Defines the layout of the Tableau Pulse metric	default is identical to the full metric page in Tableau Pulse. card is the smaller version of the metric, identical to what's found on the Pulse home page. ban includes only the value and comparison values.	No
token	The JSON web token (JWT) used after configuring a connected app	A valid JWT.	Yes

In Table 6-1, notice that a token is required when using the web component. Tableau requires developers to use and configure the connected apps functionality between the web page or application and Tableau Cloud. Configuring connected apps is best done by a web developer with support from a Tableau Cloud administrator. To learn more about connected apps and how to configure them, visit the Tableau help documentation (*https://oreil.ly/TL15o*).

Example 6-2 shows the code block structure for inclusion of a Tableau Pulse metric. You'll want to replace the src attribute with the URL of the Pulse metric being embedded, and the token with a valid JWT.

Example 6-2. HTML of a Tableau Pulse web component

```
<tableau-pulse id="tableauPulse"
 src='https://Tableau-Pod.online.tableau.com/site/mysite/pulse/metrics/metric-
 id' token='CAtoken'>
</tableau-pulse>
```

Tableau REST API

The last set of tools available to developers is the Tableau REST API. This API allows developers to programmatically complete a variety of actions in Tableau Pulse. These actions apply to metric definitions, metrics, metric insights, and metric subscriptions (followed metrics).

For metric definitions, the API includes methods to do the following:

- Create, update, and delete definitions
- Get a list of metric definitions for a Tableau Cloud site
- Get a list of metrics based on a metric definition
- Get a specified batch of metric definitions

For metrics, the API includes methods to do the following:

- Get the details of a metric
- Get a metric if it exists, or create it if it doesn't yet exist
- Create, update, or delete a metric
- Get a specified batch of metrics

For metric insights, the API includes methods to do the following:

- Generate a basic insight bundle for a metric
- Generate a springboard insight bundle for a metric
- Generate a detail insight bundle for a metric

For metric subscriptions (followed metrics), the API includes methods to do the following:

- Get the details of a subscription to a metric
- Get a list of subscriptions to a metric for a user
- Create or delete a subscription to a metric
- Update the followers of a metric
- Get a specified batch of subscriptions to a metric

- Create a subscription for a batch of users or groups to a metric

- Get a count of followers for a specified batch of subscriptions to a metric

The robustness of this API gives a developer access to all the same functionality you've seen demonstrated in the Tableau Pulse UI throughout Chapters 1 through 4. Tableau's documentation (*https://oreil.ly/h0WLZ*) on the Tableau REST API includes methods and responses for Pulse.

Summary

This chapter exposed you to advanced integrations and developer tools, allowing you to extend the reach of Tableau Pulse. Here are some key takeaways:

- Integration between Slack and Tableau Cloud requires administrators for both tools.

- You can embed Tableau Pulse summaries and metric pages inside Salesforce CRM.

- To embed Tableau Pulse inside Salesforce, you'll want to set up Tableau Cloud as a trusted URL and configure Salesforce as a connected app with Tableau Cloud.

- Tableau Pulse pages can be added as a component on a lightning page in Salesforce, either as a standalone page or within other preexisting pages.

- Tableau Pulse metrics can be embedded into web pages or custom web applications by using the Embedding API v3.

- The Embedding API v3 includes a web component for Tableau Pulse. This reduces the amount of coding necessary to include it in a web page.

- The Tableau Pulse web component can be customized with several optional properties, like height, width, and layout.

- For more programmatic control to Tableau Pulse, developers can take advantage of the Tableau REST API.

- The Tableau REST API includes methods that cover all the actions that can be done via the Tableau UI.

In the next chapter, the focus shifts from Tableau Pulse to Tableau Agent. You'll learn how to use this AI assistant to create visualizations, calculations, and data asset descriptions.

CHAPTER 7

Tableau Agent

Now that you've learned about all the facets of Tableau Pulse, it's time to move on to the other AI tool available in the Tableau platform: Tableau Agent. As mentioned in the Preface, Tableau Agent (formerly Einstein Copilot) is an AI assistant available to those authoring visualizations or Prep workflows on the web. It also works directly with the Tableau Data Catalog feature to generate summaries of data assets and content. With Tableau Agent, you can use conversant AI to build charts and calculations, receive suggestions on what to analyze, and minimize administration time. Let's dive in.

Prerequisites

Tableau Agent is currently available only on Tableau Cloud for those who have a Tableau+ subscription. Introduced in the summer of 2024, Tableau+ is marketed as an enterprise-level solution of the Tableau platform (Tableau Cloud only) with premium AI features. This version of the product line includes everything you're accustomed to with a normal Tableau license, plus the inclusion of the Data Management and Advanced Management add-ons, access to its eLearning library, an instance of Data Cloud, Tableau Agent, additional capabilities for Tableau Pulse, Data Connect service, and premium support. Unsurprisingly, this offering comes with a hefty new price tag, which is roughly 3x the traditional licensing costs for each license type.

To enable Tableau Agent, organizations must have a Data Cloud instance and connect it with their Tableau Cloud instance. Additionally, they must have Salesforce CRM with Einstein generative AI set up and enabled. The connection between the three platforms serves two primary purposes:

- Tracking and auditing of requests to AI services

- Utilization of AI services via a credits-based system

More information about the setup between these platforms is available in the help documentation (*https://oreil.ly/eEr0e*).

Visualization Authoring

Assuming you've got a Tableau+ license and Tableau AI setup within your Tableau Cloud environment, you'll be able to use Tableau Agent when authoring visualizations on the web. To begin, you'll want to open a new or existing workbook on Tableau Cloud. From there, you'll see an Einstein icon in the upper-right area of the authoring screen, next to the Data Guide icon, as seen in Figure 7-1.

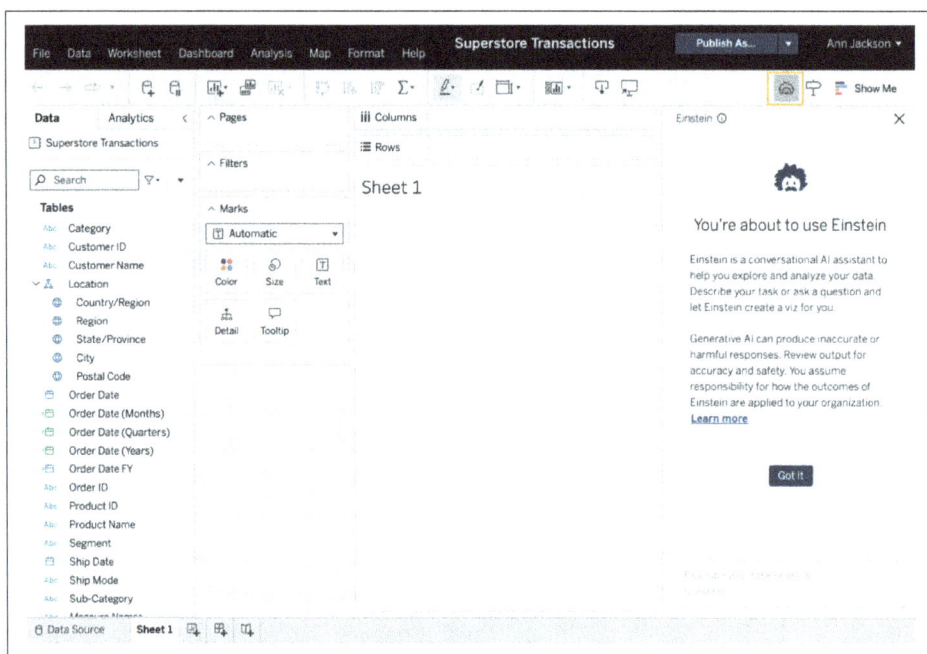

Figure 7-1. The Tableau Agent window in web authoring

Upon accessing this window and clicking through the general warning message for generative AI, you can begin interacting with it. Tableau Agent lets you know that it can assist with the following:

- Creating simple visualizations
- Creating and describing calculated fields
- Performing operations such as filtering, sorting, and grouping data
- Providing suggestions on what to analyze

You can also ask it direct questions and describe tasks you want completed. This section explores some of these features and discusses how Tableau Agent can assist. I'll be using the familiar Superstore Transactions data source first mentioned in Chapter 2.

> Although grouping is listed as a feature by the AI, Tableau documentation indicates that grouping is not currently supported.

Creating Simple Visualizations

Using Tableau Agent, you can speed up creating visualizations that may be tedious or take several clicks. I'll start with very direct examples and gradually change the language in the prompts to be more ambiguous, allowing the AI to make decisions (such as chart type) that aren't explicitly stated (see Example 7-1).

Example 7-1. Direct request for a specific chart type

Prompt

```
Make a line chart showing sales by week
```

Tableau Agent

```
I created a line chart to show sales by week. I used the 'Order Date' field
for the x-axis and aggregated it by week. The 'Sales' field is on the y-axis,
aggregated by sum.
```

Figure 7-2 shows the chart that Tableau Agent produced from the prompt in Example 7-1, which aligns with my expectations. It correctly made a line chart with the Sales field on the y-axis, trending across weeks. Now I'll ask for a specific chart type, shown in Example 7-2.

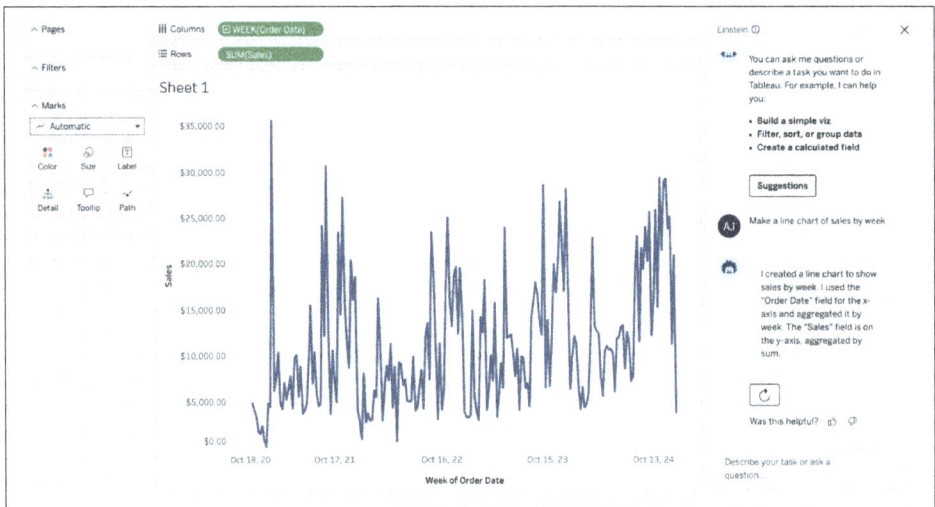

Figure 7-2. A line chart created by Tableau Agent after prompting (see Example 7-1) (see a larger version of this figure online (https://oreil.ly/lait0702))

Example 7-2. Request for a specific chart type, but with some ambiguity on the layout

Prompt

> Create a bar chart that breaks down the sales of each sub-category for each category.

Tableau Agent

> I created a bar chart that breaks down the sales of each sub-category for each category.

Figure 7-3 shows the resulting chart from the prompt in Example 7-2, which isn't exactly what I was looking for but does meet the brief stated in the prompt. I was hoping to get a bar chart with both Category and Sub-Category on rows, avoiding the repetition of empty sub-categories that clutters up the visualization.

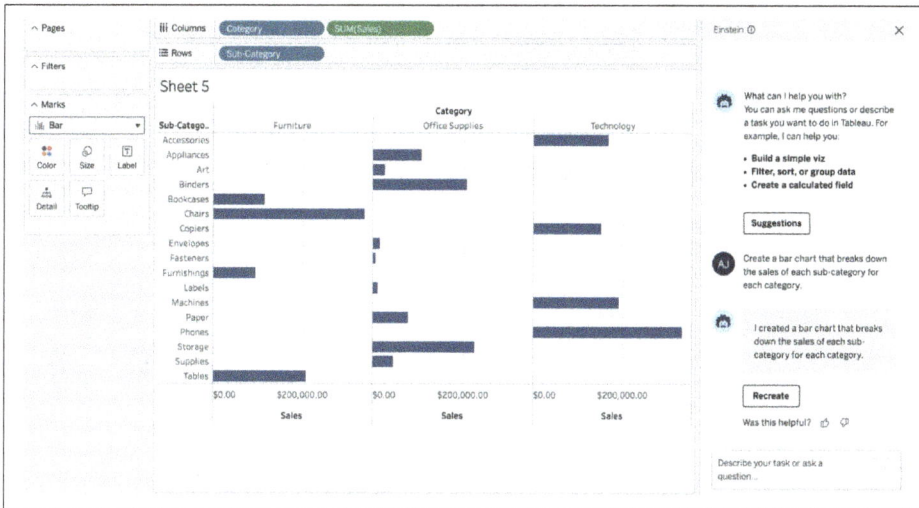

Figure 7-3. A bar chart created by Tableau Agent after prompting (see Example 7-2) (see a larger version of this figure online (https://oreil.ly/lait0703))

Since it's conversant AI, I can follow up with another prompt to try to get the chart I'm aiming for (see Example 7-3).

Example 7-3. Follow-up request to adjust the chart layout

Prompt

> I want both the category and sub-category to be on the left side

Tableau Agent

> I updated the bar chart to place both the category and sub-category on the left side, breaking down the sales for each sub-category within each category.

And the resulting chart, as shown in Figure 7-4, now aligns with my expectations. The output from this prompt reveals that the AI is in tune with the layout of the authoring screen and understands what it means if something is placed on the *left side* of a visualization.

Tableau Agent can also handle requests to see data without specifying a chart type. This can be useful if you're not sure of the type of chart that you want to create (Example 7-4).

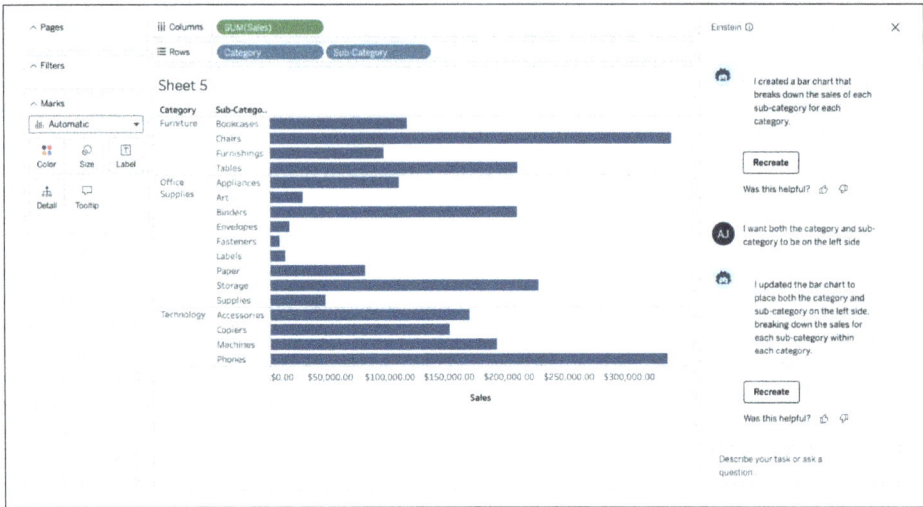

Figure 7-4. The updated bar chart after specifying placement of fields (see Example 7-3) (see a larger version of this figure online (https://oreil.ly/lait0704))

Example 7-4. Specific data elements with no description of desired chart type

Prompt

```
make a chart showing the sales and quantity for every product
```

Tableau Agent

```
I created a bar chart to show the sales and quantity for every product. The
'Product Name' is on the y-axis, and both 'Sales' and 'Quantity' are on the x-
axis.
```

Figure 7-5 shows the resulting chart, based on the prompt from Example 7-4, which includes identifying the correct fields. However, the goal I have for this chart is to see the relationship between the two values for every product. So, I'll follow up with another prompt to guide the AI toward the type of insights I'm trying to uncover, which are best shown using a scatterplot (see Example 7-5).

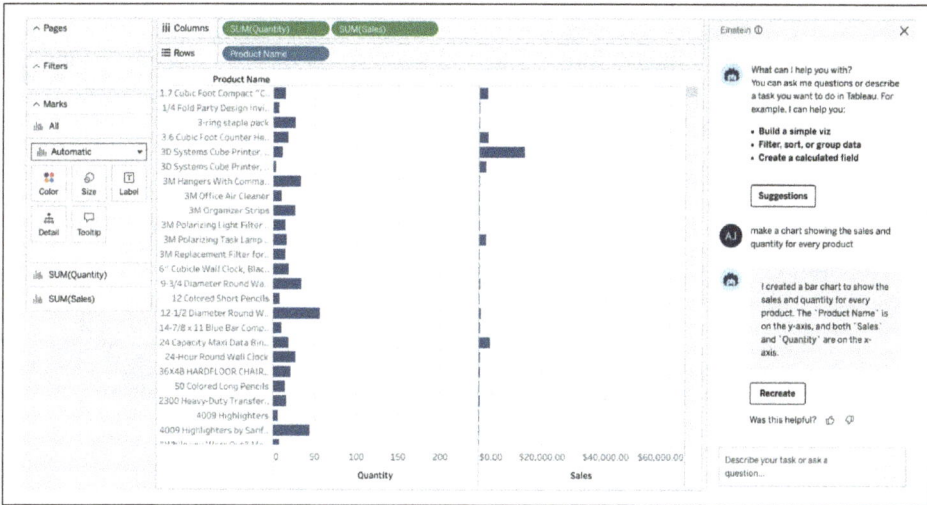

Figure 7-5. Chart created by AI when no chart type was specified in Example 7-4 (see a larger version of this figure online (https://oreil.ly/lait0705))

Example 7-5. Request to produce a chart based on a specified goal

Prompt

> I want to see the relationship between the two values for all products

Tableau Agent

> I updated the viz to show the relationship between sales and quantity for every product. I used a scatter plot to better visualize the relationship between these two values.

With the key phrase of *relationship between the two values* included in the Example 7-5 prompt, which is how Tableau documentation typically describes the purpose of scatterplots, the resulting visualization is what I was aiming for, as shown in Figure 7-6.

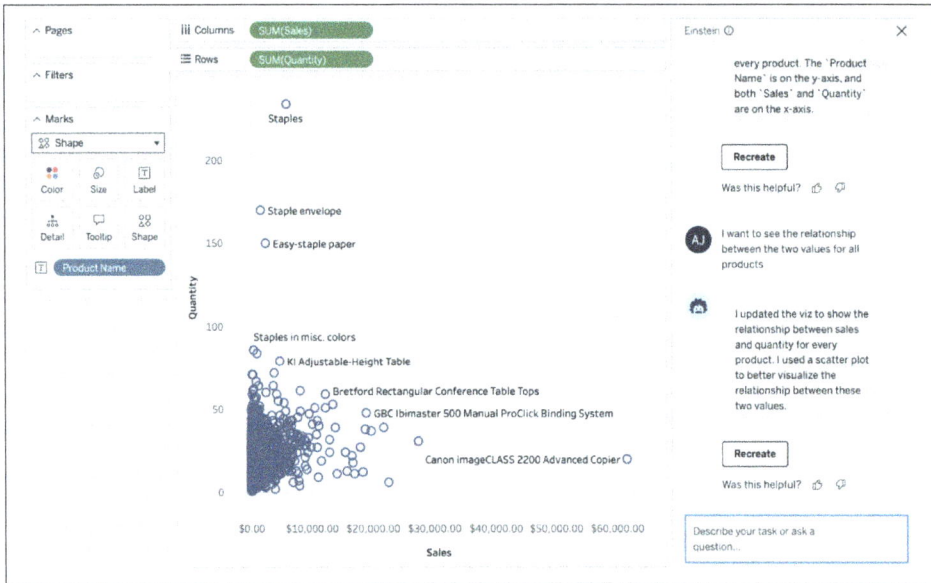

Figure 7-6. The updated chart after providing more context in the prompt of Example 7-5 (see a larger version of this figure online (https://oreil.ly/lait0706))

Knowing that the AI does better when provided more contextual information on layout or the analytical task is extremely useful. Here's one more example (Example 7-6) where there's ambiguity in the type of chart to create, but the purpose of the analysis is clear.

Example 7-6. Specific analytical task with no description of the chart type

Prompt

> make a chart showing the part-to-whole relationship of Category and Sub-
> Category when looking at the quantity

Tableau Agent

> OK. I can't make exactly what you requested, but here's a viz based on the
> fields you mentioned.

The resulting chart is a treemap, as shown in Figure 7-7, which is what I was mentally expecting. The correct fields are identified, and because of the number of unique subcategories, provides a cleaner visual than a pie chart. It's somewhat surprising that Tableau Agent mentions that it can't make exactly what I requested, since the chart completely fits the brief.

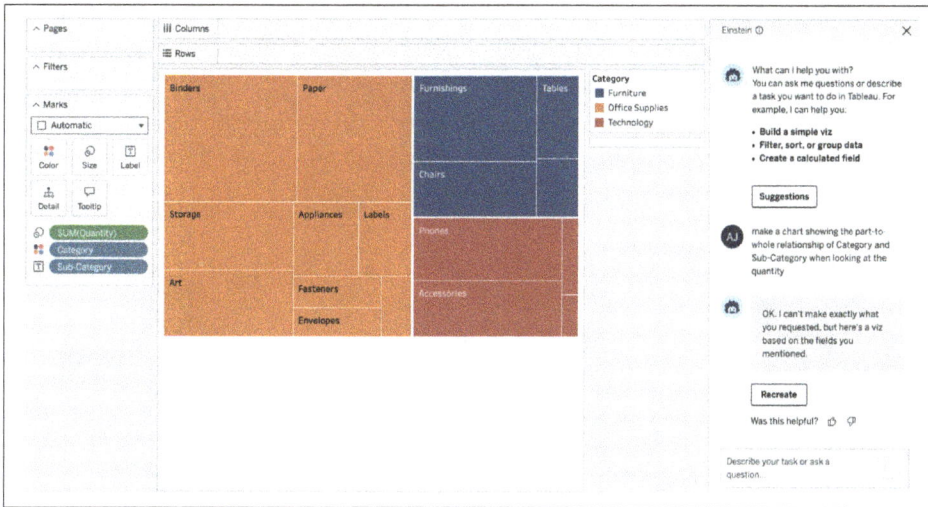

Figure 7-7. Chart created by AI after providing an analytical task in the prompt of Example 7-6 (see a larger version of this figure online (https://oreil.ly/lait0707))

Given how the AI performs when constructing visualizations, it makes sense to provide as much context as possible in your prompt. I would recommend including the chart type (if you have one in mind), any specifics you're interested in for the layout, and the type of analysis you're trying to achieve.

Working with Calculated Fields

Next, I want to go through some examples involving calculated fields. With Tableau Agent, you can create calculated fields, get a description of how they work, and receive assistance on debugging calculations that aren't working the way you want. Example 7-7 shows the prompt for a simple calculation.

Example 7-7. Simple calculation

Prompt

> Create a calculated field that is the number of days between the Order Date
> and Ship Date

Tableau Agent

> This calculation returns the number of days between the Order Date and Ship
> Date. You can use this calculated field in your Viz to analyze the shipping
> duration. Add the 'Order Date' and 'Ship Date' fields to your Viz to see the
> calculated shipping duration.

The prompt instantly created the exact calculation I was looking for and gave it a friendly name—definitely a huge time-saver. Figure 7-8 shows the result, which includes a modal pop-up of the calculated field for review.

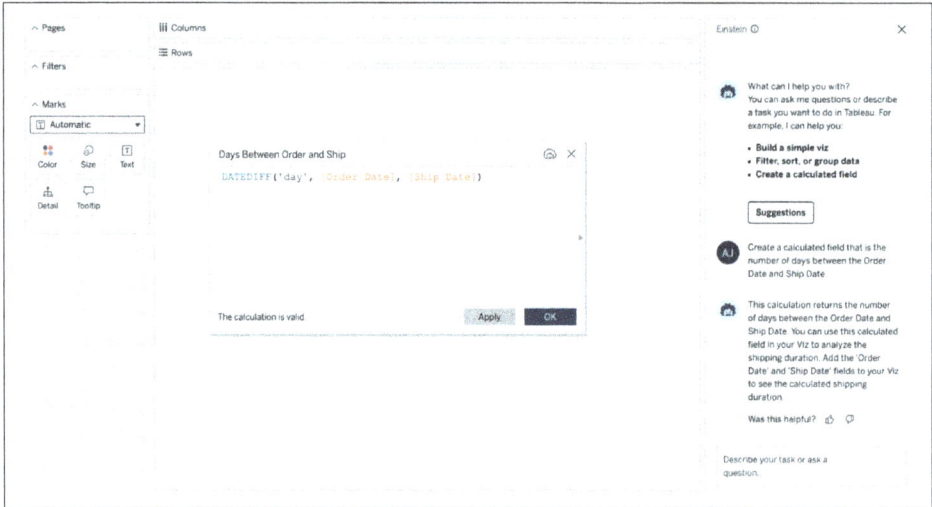

Figure 7-8. Calculated field made based on the prompt in Example 7-7 (see a larger version of this figure online (https://oreil.ly/lait0708))

Now I want to get a little more sophisticated and see if Tableau Agent can create a calculated field that uses more advanced elements, like table calculations. Since table calculations rely on the layout of the data in the visualization, I'll begin with a chart that includes the right elements but doesn't quite have the calculation I want (see Example 7-8).

Example 7-8. Advanced calculation

Prompt

> Create a calculated field showing the percentage difference between each month for sales.

Tableau Agent

> This calculation computes the percentage difference in sales between each month. To use this calculation, place the 'Order Date (Months)' field on the Columns shelf and the new calculated field on the Rows shelf in your Viz. Then, set the table calculation to compute using 'Order Date (Months)'.

Figure 7-9 shows the result, which includes a friendly name for the calculation and accurately uses the LOOKUP() function, which is a table calculation. It also includes how to specify the compute direction of the table calculation. Interestingly, it also includes an optional `IFNULL` function in the calculation, returning null or empty values as 0.

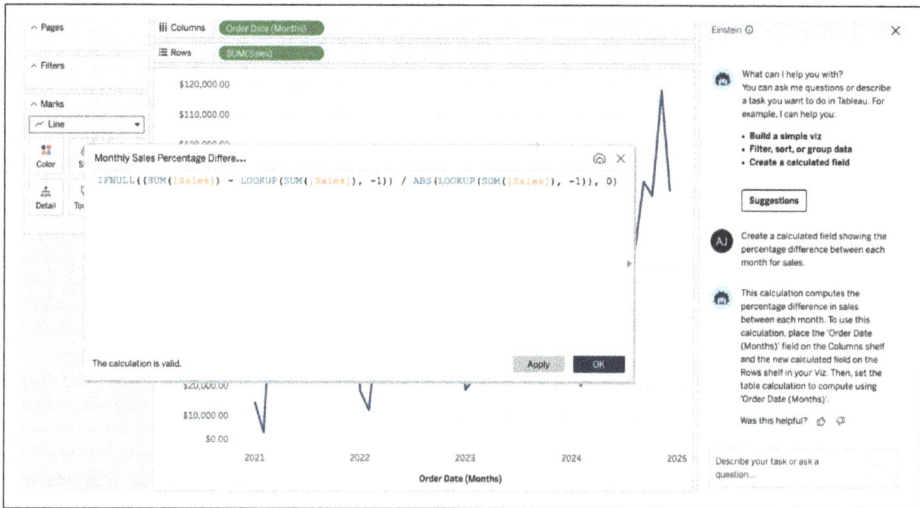

Figure 7-9. Advanced calculation made after the prompt in Example 7-8 (see a larger version of this figure online (https://oreil.ly/lait0709))

Finally, I want to try a compound concept within a calculated field, one that requires both an LOD expression and a function. This would be the type of calculation that requires some time to conceptualize and requires the creator to likely build out the calculation in multiple steps while validating its functionality along the way (see Example 7-9).

Example 7-9. Multiple calculation concepts

Prompt

> For every state, find the very first order date using customers whose first name is exactly 5 characters

Tableau Agent

> This calculation finds the very first order date for each state using customers whose first name is exactly 5 characters. Add the 'State/Province' and 'First Order Date for 5-Character First Names' fields to your Viz to display the results.

Figure 7-10 shows the calculation constructed, and I have to admit this output is impressive. Tableau Agent used string calculations that would likely require a bit of trial and error and correctly combined it with an LOD expression to isolate the value by State/Province. Although the request is pretty esoteric, asking the AI to construct the calculation was much more efficient than attempting to create the calculation myself.

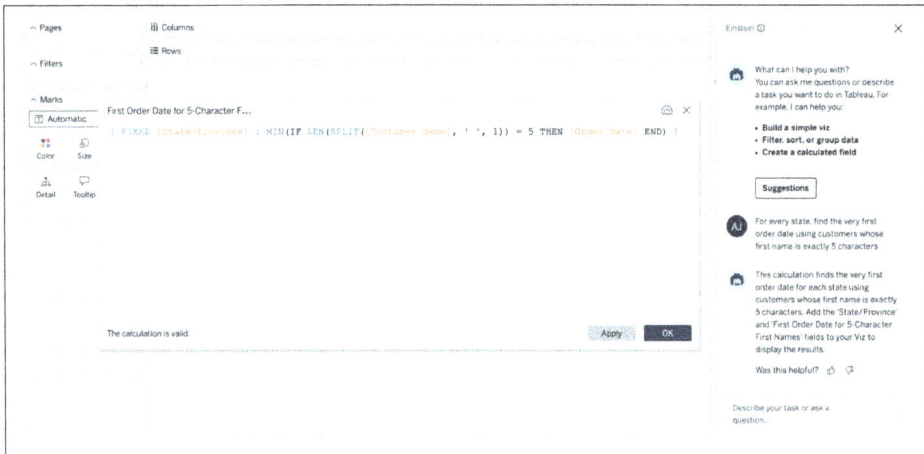

Figure 7-10. A complex calculation made by AI from the prompt in Example 7-9 (see a larger version of this figure online (https://oreil.ly/lait0710))

Working with calculations may be the biggest strength of Tableau Agent. Let's see how it describes a calculation and see if I can get assistance in modifying it to meet my needs.

Notice in Figure 7-10 that there is a small Einstein icon in the calculated field window. Clicking this icon causes Tableau Agent to list how it can assist with calculations. First, I'll ask it to explain the calculation Profit Rank, which I've created to rank the Profit value for each sub-category (see Example 7-10).

Example 7-10. Explain and debug the existing calculated field

Prompt

    Explain Profit Rank

Tableau Agent

    The 'Profit Rank' calculated field assigns a unique rank to each item based
    on the total profit, with the highest profit receiving rank 1. It uses the
    RANK_UNIQUE function to ensure that each rank is distinct, and it sorts the
    profits in descending order.

Figure 7-11 shows the visualization with the Profit Rank field I've constructed and the explanation.

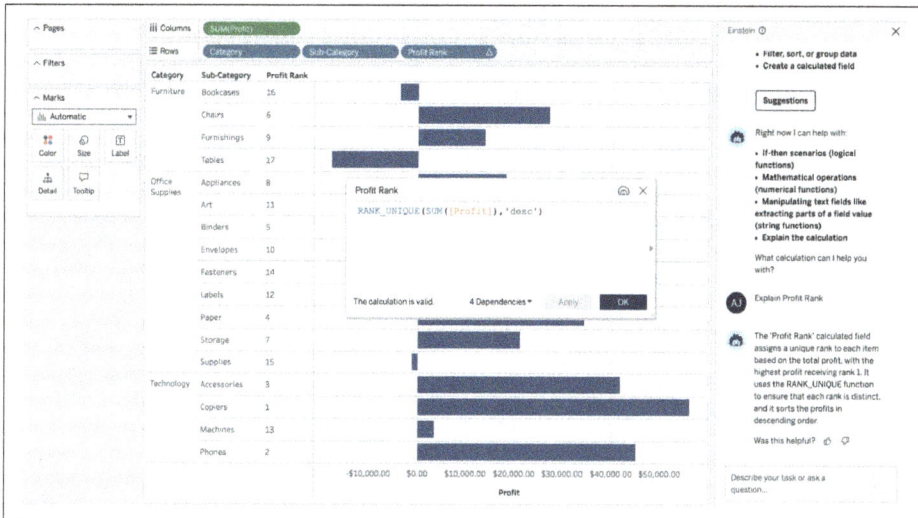

Figure 7-11. Profit Rank calculation explained from the prompt in Example 7-10 (see a larger version of this figure online (https://oreil.ly/lait0711))

You'll notice within the visualization that the rank runs across all the sub-categories, ignoring the Category field. But what I was hoping to achieve with this calculation is for each sub-category to receive a rank within its parent category. Let's see whether Tableau Agent can assist (see Example 7-11).

Example 7-11. Request to adjust how rank is computed

Prompt

> How can I change the calculation so that the rank starts over for every Category?

Tableau Agent

> This calculation ranks the sum of profits uniquely within each category. To use this in your Viz, place 'Category' on the Rows or Columns shelf, 'Sub-Category' on the Rows shelf, and 'Profit Rank' on the Text or Label shelf. Set the 'Compute Using' setting to 'Sub-Category' to ensure the ranking restarts for each category.

This response includes instructions about how to place the fields on the canvas but, more importantly, it does correctly identify how to modify the table calculation to restart the ranking for every Category. Figure 7-12 shows the final visualization, after taking Tableau Agent's advice and setting the Compute Using to Sub-Category.

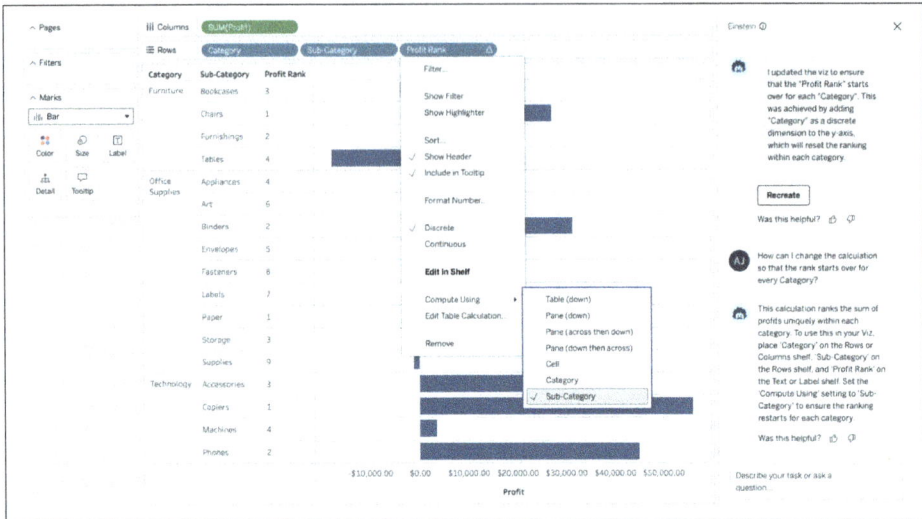

Figure 7-12. Updating the Compute Using method of a table calculation from the prompt in Example 7-11 (see a larger version of this figure online (https://oreil.ly/lait0712))

Filtering

In addition to assisting with visualization and calculation construction, you can use Tableau Agent to limit data within visualizations. This section explores this feature to see how it can be used to streamline these types of tasks, going from simple concepts to more complex scenarios.

First I want to see what happens when I ask for very straightforward date filtering. I've already set up a line chart showing Sales by Month for each Segment as the chart for these prompts (see Example 7-12).

Example 7-12. Simple date filter

Prompt

    Limit the data to the last 18 months

Tableau Agent

    I updated the viz to limit the data to the last 18 months.

As expected, the AI placed the Order Date field on the Filters shelf. It also exposed a relative date filter, which is helpful, since a few types of date filter designs can be used. Figure 7-13 shows the result.

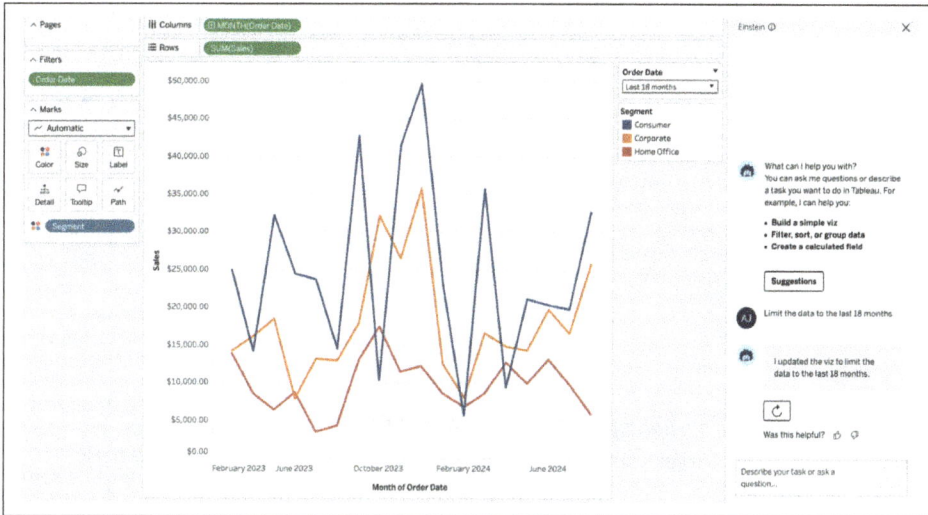

Figure 7-13. Date filter created by AI from the prompt in Example 7-12 (see a larger version of this figure online (https://oreil.ly/lait0713))

Next, I want to try a filter based on a dimension. I'll stick with the same chart and use Segment as my target dimension for filtering (see Example 7-13).

Example 7-13. Filter by a dimension

Prompt

    I only want to see Consumer and Corporate in the chart

Tableau Agent

    I updated the viz to only include the 'Consumer' and 'Corporate' segments in
    the chart.

This time I intentionally chose to focus on my desired outcome, as opposed to describing the action to take. The AI correctly resolves the task and adds a Segment filter to the Filters shelf, as shown in Figure 7-14.

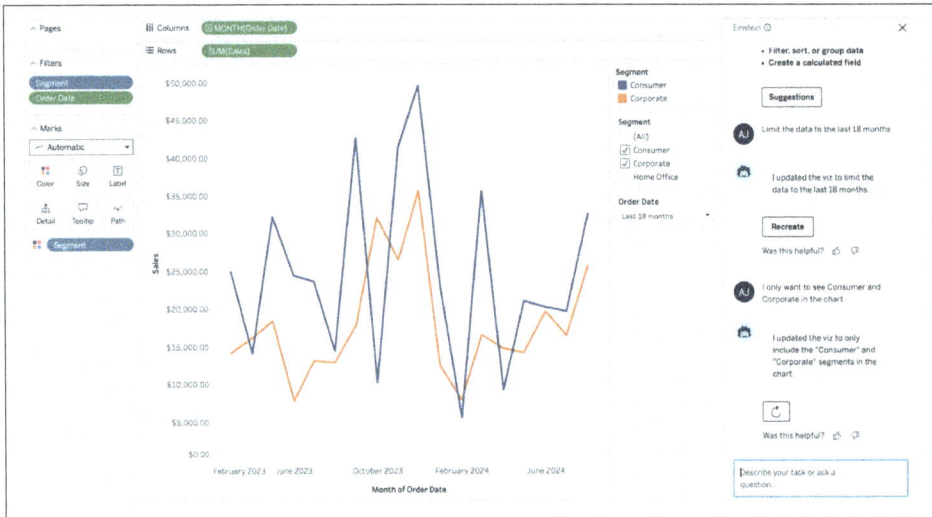

Figure 7-14. Updated visualization with a Segment filter from the prompt in Example 7-13 (see a larger version of this figure online (https://oreil.ly/lait0714))

Now let's see if the AI can keep up with a filter operation based on a numerical value (see Example 7-14).

Example 7-14. Filter by a measure or numerical value

Prompt

> Only include rows of data where the sales value is more than 100

Tableau Agent

> I updated the viz to only include rows where the sales value is more than 100.

Tableau Agent correctly interprets my prompt and adds a filter for Sales to the Filters shelf, as shown in Figure 7-15.

> This is where the AI reaches its limit in terms of interpretation and correct action. To get the result in Figure 7-15, I had to iterate through several differently worded prompts. Each unsuccessful prompt resulted in the AI adding an aggregated version, SUM(Sales), as a filter, as opposed to the individual value. The key to getting to this result was the very explicit indication of *rows of data*, which provided the context necessary to look at the individual Sales values instead of the sum.

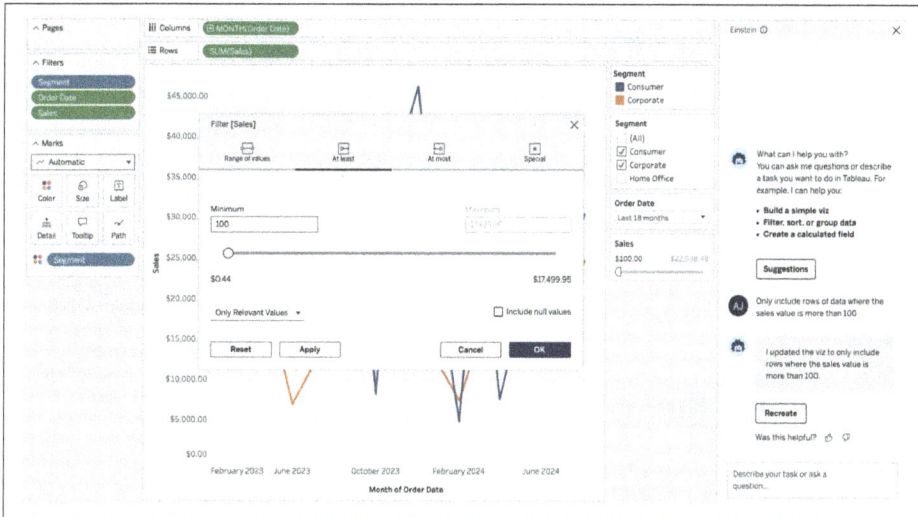

Figure 7-15. AI-generated filter based on a numerical value from the prompt in Example 7-14 (see a larger version of this figure online (https://oreil.ly/lait0715))

Following Best Practices for Prompting

As you saw throughout the preceding examples, phrasing of your prompts is crucial. Here are some best practices for working with Tableau Agent while authoring:

Use direct and explicit language
   Tableau Agent works best when you use imperative verbs to tell it specifically what you want. *Make*, *create*, *build*, *write*, *explain*, and *show me* are all great prompt starters.

Minimize irrelevant data
   As with all good analysis, it's best if you hide or exclude unnecessary fields in your data. This eliminates the opportunity for the AI to offer suggestions based on superfluous information or similarly named fields.

Use Tableau language
   If you want a calculation to be built using a specific function, name it in your prompt. Similarly, use contextual phrases that frequently appear in documentation to guide the AI.

Work in steps
   The AI functions best when it's given one direct action to follow through on. Break these tasks into separate prompts and iterate through them.

Ask questions

> If you're not exactly sure what visualization or calculation you're trying to create, phrase your prompt in the form of a question. Start with words such as *how*, *what*, and *which*.

In addition to these basic best practices, don't be afraid to provide feedback on whether the AI output was helpful or not by using the familiar thumbs up/down icons. Remember, the LLM doesn't store or directly learn from your prompt because of the Einstein Trust Layer (refer to Chapter 2), so you won't see an immediate improvement. However, your feedback is likely aggregated and used to enhance the overall model.

Tableau Prep Builder

Tableau Agent is also available in the Add Field editor within Prep Builder on the web. You'll notice that once you're in the editor, there's a section for Tableau Agent (Einstein), as shown in Figure 7-16.

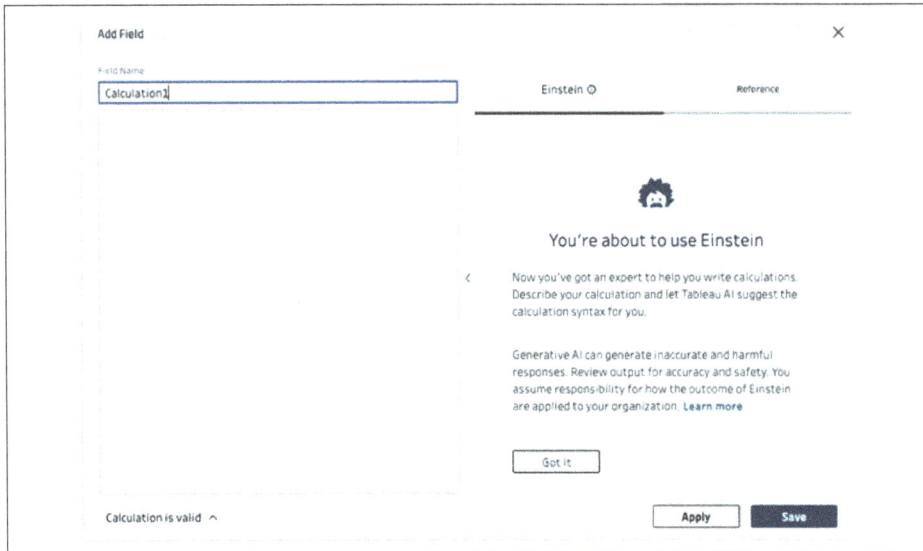

Figure 7-16. Calculation editor with Tableau Agent in Prep

The experience with Tableau Agent in Prep is virtually identical to that of the web-authoring support you receive when working with calculations. In general, you'll be able to use the functions and operators that are available within the Reference tab of the editor. These include logical operators, fixed LODs, numerical aggregations, and more. In particular, Tableau Agent in Prep Builder also includes the ability to construct regular expressions.

Regular expressions (regex) are patterns used to match character expressions in string fields. They are powerful, but often identifying the correct syntax to isolate the pattern is challenging. This is where Tableau Agent can help. Example 7-15 shows what happens when asked to create a calculation that parses out the four-digit year in the Order ID field, which includes a two-letter country code, four-digit year, and a six-digit order number separated by dashes (example: CA-2021-100867).

Example 7-15. Constructing a calculation with a regular expression

Prompt

> Write a regular expression to parse the four digit year from the Order ID
> into its own field.

Tableau Agent

> This calculation uses a regular expression to extract the first sequence of
> four digits from the 'Order ID' field, which represents the four-digit year.

Figure 7-17 shows the Add Field editor and conversation with Tableau Agent. In addition to constructing the calculation, Tableau Agent also provides a button to replace the calculation in the editor with the regular expression suggested.

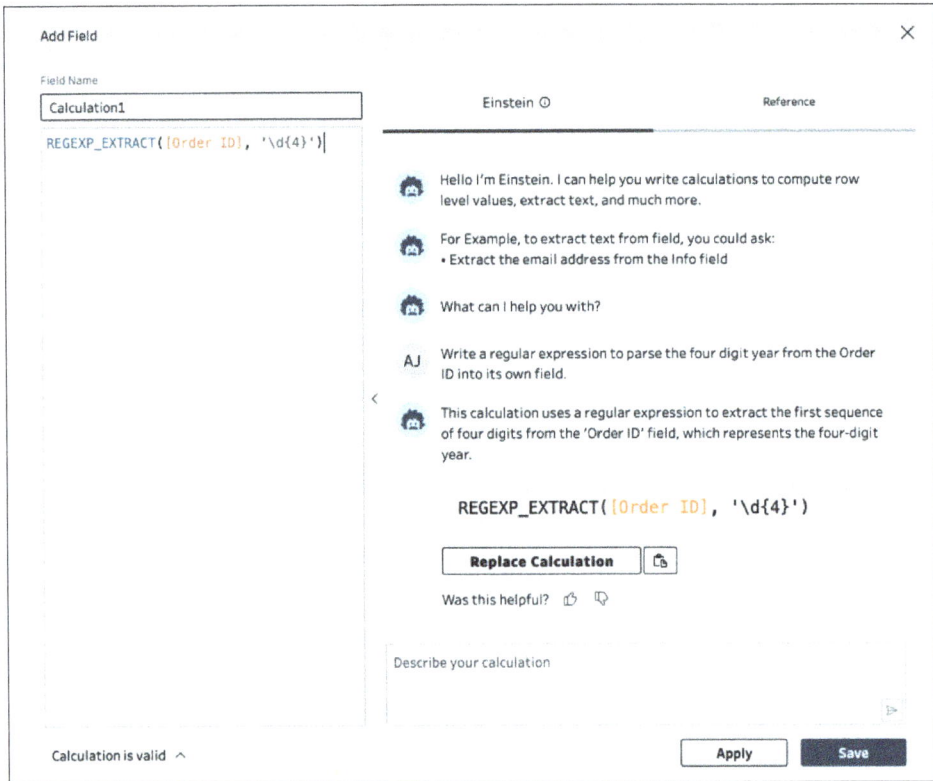

Figure 7-17. Regular expression created based on the prompt in Example 7-15 (see a larger version of this figure online (https://oreil.ly/lait0717))

> Separate workflow steps, like joining or unioning data, pivoting and aggregation, creating new rows, predictions, inserting a script, and output, are not currently supported. Similarly, there is no consultative guidance or explanations available for these steps when using Tableau Agent.

Tableau Data Catalog

Finally, Tableau Agent is available for usage with the Tableau Data Catalog. When writing descriptions for workbooks, data sources, or tables, you'll notice a Draft with Einstein button available. Figure 7-18 shows how it appears when editing a description for a table.

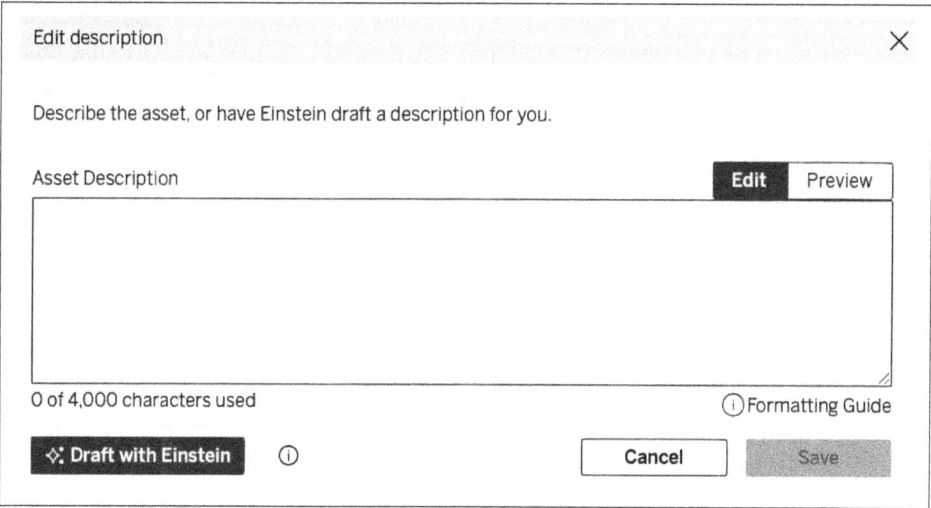

Figure 7-18. Draft with Einstein option for describing a table

The AI will rely on metadata like field and table names to generate a description. Descriptions can be regenerated by clicking the Draft with Einstein button again. You can also modify the automatically generated description and add formatting as desired. Figure 7-19 shows a description generated for the credit card transactions data set used in the finance example from Chapter 5 (refer to Figure 5-9).

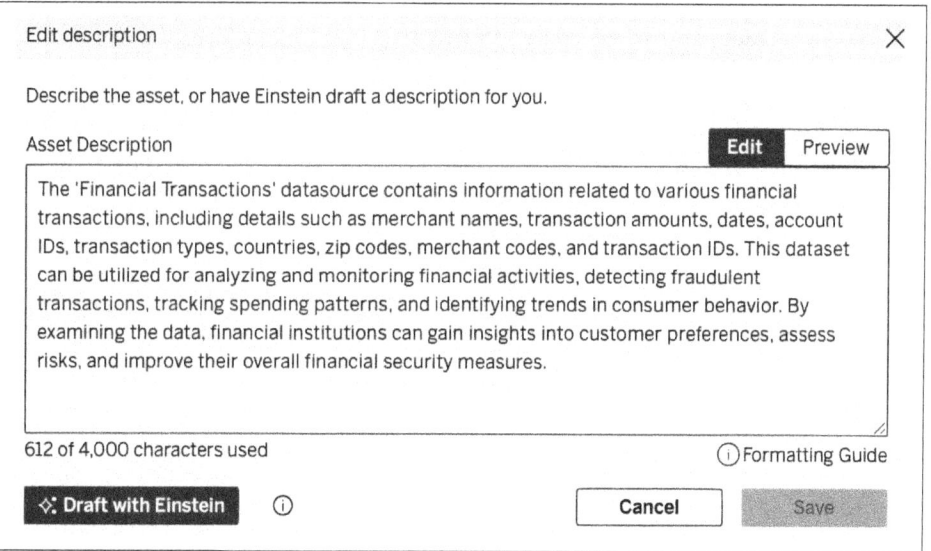

Figure 7-19. A published data source description generated using Draft with Einstein

As you can see, the AI is able to pull out the field names within the published data source and suggest how it can be analyzed, saving a ton of time and avoiding many typos in the process.

Summary

This chapter walked you through many examples of working with Tableau Agent throughout the Tableau platform. You've seen how it can assist with visualization authoring and calculation creation. Here are some key takeaways:

- Tableau Agent can make simple visualizations and is most effective when describing the analytical task or chart type you want.
- Tableau Agent can construct a variety of calculated fields, from the use of simple functions to complex calculations that involve a mixture of concepts.
- Tableau Agent can filter data within a worksheet, but will likely need explicit instructions when filtering numerical values.
- You'll have the most success with Tableau Agent if you give it explicit instructions. Begin your prompts with action verbs if you know what you want. If you're unsure of the outcome you're trying to achieve, formulate your request as a question.
- Tableau Agent within Tableau Prep is currently limited to calculation creation and assistance.
- Tableau Agent can use metadata information to automatically generate descriptions for workbooks, data sources, and tables.

Chapter 8, the final chapter, discusses the influence these new AI technologies have on the platform and previews what new features may be coming down the pipeline.

What's Next?

You've made it through the breadth of AI tools currently available in Tableau, and you now know how to maximize them to support analytics at your organization. However, that's just the beginning. The use of AI to enhance productivity and insights is still in its early stages. Therefore, this chapter aims to present my observations and perspectives on what the future might hold. I'll discuss the future of the Tableau product line, consider the direction AI is going with analytics, and give you a broad overview of the larger AI landscape. Much like Salesforce's own forward-looking statement cautioning stockholders not to make purchasing decisions based on the content of their technical presentations, don't bank any key decisions on what's discussed in this chapter; wait for an official product release before taking action.

The Future of Tableau

Since the acquisition of Tableau Software by Salesforce, there's been an undercurrent of belief that Salesforce will eliminate non-cloud-based products from the Tableau ecosystem. And with the introduction of AI tools, the company has all but solidified this direction. At Salesforce's annual customer conference, Dreamforce (September 2024), Tableau President and CEO Ryan Aytay debuted Tableau Einstein, which is Tableau code rewritten to function on Salesforce's (cloud-only) Core platform.

Tableau Einstein is a reimagined Tableau experience that aims to finally allow organizations to access and leverage all their data, where it already resides. It is also described as a full analytics solution, one where data modeling, semantic modeling, metric creation, and visualizations can all reside in a strongly governed environment. The center of this technology is Data Cloud, which offers zero-copy connectivity to all major cloud data warehouse solutions on the market. A new concept announced with Tableau Einstein is the idea of *workspaces*, a place to organize, collect, govern, and share analytics with the right audiences (both internally and externally). A workspace can be anything from a personal playground to an internal team space to an external marketplace. Tableau Einstein features all the AI tools discussed in this book, packaged in a slightly different way.

Tableau Einstein also brings in a new UI for Tableau. Working in Tableau, Einstein is eerily similar to the tabbed pages and componentized assets you may be familiar with in Salesforce's other products. And this concept speaks to one of the key messages Tableau has been championing: analytics in the flow of work. With this new UI, the new VizQL Data Service, and a marketplace to share and reuse analytical assets, the next generation of Tableau is aimed at weaving visualizations and data-driven insights into every aspect of a user's workflow. Although there is still a space for dashboarding, the hyper focus on personalization and contextualization points to serving up bite-sized information at just the right time.

> At time of writing, Tableau Einstein has not yet been released. You can watch a video of the Tableau Einstein announcement at the Dreamforce 2024 Tableau Keynote on Tableau's website (*https://oreil.ly/6-nxE*).

Tableau Pulse

I've had the benefit of tracking the initial release of Tableau Pulse and subsequent releases during the writing of this book. So to me, it seems fairly clear which direction Tableau Pulse is headed. There's been a significant amount of progress to ensure that metrics are more integrated into the ecosystem, specifically with the newly announced inclusion of metric definitions in the data catalog feature and the ability to embed metric objects in dashboards.

Additionally, there's been a recent push to add new functionality like manual goals for metrics, advanced sorting of metrics, and expanded AI reach between metrics as premium Tableau+ features. As Tableau Pulse continues to evolve, we'll likely see most additional advanced or new features released as premium-only functionality.

Some big gaps still remain in Tableau Pulse that I am hoping will be filled in. Chief among them is the necessity for all metrics to have a time dimension associated with them. While time is very commonly used in analysis, requiring a time dimension within a visualization becomes a limiting factor in many scenarios. I'd like to see the expansion of chart types and for time dimensions to be used as only a filter.

I'd also like to see the functionality of Pulse expand such that separate metrics function off the same time horizons and dimension filters—basically a reimagined version of a dashboard, where related metrics can all be manipulated based on global filters. Although this is somewhat accomplished in the breakdown section of metrics, expanding the functionality to tune several metrics to the same filtered cut of data seems inevitable to both avoid user confusion and enhance the summarized insights served up. Similarly, it makes sense to start incorporating comparisons and correlations among different metrics as a way to find their hidden relationships.

Though I'm not sure if it's likely in the very near future, the final aspect I'd like to see enhanced is the ability to customize the look of metrics. Current customization and styling features are nearly nonexistent. Aside from the ability to include color indicators when the *value going up* is specified, there isn't a meaningful way to customize charts with colors, labels, shapes, or amount of detail. In particular, charts displayed on the mobile version can lack specificity to see the range and magnitude of a value, instead favoring the look of a sparkline, which is meant to quickly convey the shape or trend of the data but is not useful for identifying exact values. And if the future involves integrating Pulse metrics into the flow of work, it would follow that they should be seamlessly integrated aesthetically.

Tableau Agent

Tableau and Salesforce definitely aren't shying away from the concept of AI assistance anytime soon. In fact, it's fair to say that they're doubling down on this concept and envisioning a future where AI acts not only as an assistant, but also autonomously. This aligns with Tableau's latest branding message of being an "AI-powered analytics platform." So in the near term, AI assistance offered by Tableau Agent is primed to expand quickly.

While still in the works, Tableau has already mentioned Tableau Agent for building dashboards. I imagine this to be the creation of multiple charts at the same time based on a prompt that is more analytically oriented (e.g., *Make a dashboard for me to track and share quarterly sales performance*). Perhaps this feature will eliminate the need to use individual worksheets and instead begin the authoring experience on the dashboard canvas.

Tableau has also mentioned that future versions of Tableau Agent will be able to take on a more consultative role. This means expanding functionality from purely task-oriented assistance to the inclusion of visualization best practices, recommendations on chart types, and potential inclusion of common methods for analyzing certain types of data (like customer data). I've also heard mention of *retrieval-augmented generation* (RAG). RAG is an enhancement to LLMs that allows model builders to incorporate trusted external sources for more detailed information, particularly when the subject is extremely technical in nature. So, there may be a future where the richness of content created by subject matter experts on the internet is included and served up via Tableau Agent.

I would like to see task-oriented assistance expand much further in capability. Currently, Tableau Agent constructs calculations and moves fields around on the various worksheet shelves but doesn't offset the tediousness of formatting or designing.

Including this type of functionality could really speed up creation and minimize the painstaking task of ensuring uniformity within your charts and dashboards. The simplest example would be the ability to specify a font and hierarchy of font sizes across all elements; that could save a tremendous amount of time and effort.

Other administrative tasks are also poised to use Tableau Agent. Everything on the content management side could benefit from AI performing bulk operations like generating descriptions, tagging content, deleting stale content, surfacing redundant content, specifying user permissions, and more. In a future where governance seems to be a much more prominent theme, giving administrators the ability to perform these types of operations in a no-code way at scale will be invaluable.

Finally, I think we'll see Tableau Agent expand further into the data-preparation space. Often thought to be the toughest part of analytics, data preparation includes the task of finding, cleaning, combining, and shaping data sets for use. If Tableau Agent can take language-based requests that would traditionally rely on advanced query or coding capabilities, organizations and analysts will be able to extract more value out of a larger portion of the data they're capturing.

AI and Analytics

As discussed in Chapter 1, other analytics and BI platforms appear to be on a similar path as Tableau: relying heavily on AI to improve productivity, reduce the barrier of entry for constructing analytical content, and automate the surfacing of data-driven insights to take action. This means that seasoned analytics professionals are poised to start training, tuning, and policing AI created or assisted content.

Traditional roles of data professionals like data engineer, analyst, and data scientist could turn into data aggregator, prompt creator, and model trainer, respectively. As an example, the data engineer may no longer need to rely on their expansive technical knowledge of accessing, ingesting, and organizing data sources, but instead will need to focus on the task of ensuring that AI has access to all the organization's data. This could involve more intensive governance work and ingesting *dark data*—that is, data that isn't currently utilized in a meaningful way within the organization—to expand AI's knowledge.

The future analyst may not spend as much time answering questions or building data products for business users, but instead could craft semantic models based on areas of the business and specify guidelines for creating analytical prompts. And finally, the data scientist, who has spent time greenfielding analytical models to solve problems, may lose some of the creativity and innovation that was previously a key component of their role. Instead of ideating on new models themselves, they will likely have to scrutinize, maintain, and refine models served up by AI (or the AI models themselves) and training data as it expands into more facets of business.

While this future of more access to data and more meaningful analysis sounds idyllic, the path there has pain points that are similar to those that have existed in analytics for the past 20+ years. First and foremost is the rationalization and collecting of all organizational data. This moving target has gone through a massive number of iterations and solutions, from storage solutions like Hadoop to cloud data warehouses like Snowflake, giving the feeling that data projects are perpetually in a state of migration to the latest and greatest platforms. AI will add a new layer to this, hopefully one that will finally see governance, curation, and organization of data mature.

With the support and assistance that AI can bring to working with data, I'm hopeful it's only a matter of time before the focus shifts back to utilizing the contents of organizational data instead of being a perpetual game of data systems migration. Similarly, AI will hopefully allow for a real expansion of analytical thinking to the everyday knowledge worker and enable a true increase in taking action based on data. In a sense, I'm talking about getting back to the basics of what analytics is all about: better understanding and informed decision making.

The Future of AI

There's no doubt in the immediate term that AI is going to continue to expand its reach, particularly in the business world. We're in a transitional phase right now where, to enhance the effectiveness of AI, more knowledge will have to be poured into the models. Once this phase of consumption matures, we'll likely begin a phase of refinement and increased AI autonomy. AI autonomy will mean that low-complexity tasks, interactions, and requests are handled without any human participation.

Additionally, specialization will likely increase as AI continues to progress. Much like the app-for-everything culture we have today, there will likely be an AI model for narrowly focused subject areas. Time will tell whether these new models will be built from the ground up or in-house, (much like the custom software solution trend of the early 2000s) or if AI organizations like OpenAI will provide off-the-shelf solutions that can be tuned to meet business needs (while still maintaining data privacy).

In the nonbusiness realm, the increase in AI should also cause an increase in individual creativity and expression. Content creation, like generative art and video, is already becoming extremely popular, and that trend is poised to expand as social media organizations like Meta and X fold AI assistants into their platforms.

And finally, with the release of Apple's first generation of integrated AI tools within its iPhone iOS 18 operating system in the last quarter of 2024, AI assistance in the form of internet searching, personal administrative tasks, and information aggregation will undoubtedly become mainstream.

Summary

Although this chapter has been mostly speculative, I hope it has given you a sense of what's to come with not only Tableau AI tools, but with AI in both analytics and mainstream applications. Here's what you can likely expect:

- Tableau to become a cloud-first or cloud-only platform with a new modern UI
- Tableau Pulse to receive more features designed to proactively alert and serve up the right information at the right time to end users
- Tableau Agent to expand its capabilities to include both consultative support and time-intense administrative (click-heavy) tasks
- AI features in the Tableau and Salesforce ecosystem to be sold at a premium
- A push for data governance, cataloging, and curation to enhance the knowledge of AI models
- A transition in the key responsibilities of data professionals, from the current technical hands-on keyboard roles to those that are more consultative, governed, and focused on refinement
- AI to become more autonomous, in both analytics and day-to-day interactions
- AI to expand its role in mainstream life, by way of content creation support, information aggregation, and assistance with mundane tasks

Congratulations on reaching the end of the book!

I hope you had as much fun reading it as I did writing it. Along the way, I trust you've gained a solid understanding of the AI tools available in Tableau and how to use them effectively. I'm excited to see how you leverage these tools in your own data projects.

Feel free to reach out via email at *tableauAI@jacksontwo.com* or connect with me on social media.

Index

supervised machine learning, 2

T

Tableau
    competitors to, AI in, 11-12
    future of, 147-151
    history of, vii-x
Tableau Agent
    ambiguity in prompts, 128, 132
    data professional's responsibilities when
        using, 9
    future of, 149
    in Prep Builder, 142-144
    LLMs in, 3
    overview of, xi, 125
    prerequisites for using, 125
    using with Tableau Data Catalog, 144
    visualization authoring using
        best practices, 141
        creating simple visualizations, 127-133
        filtering data, 138-140
        overview of, 126
        working with calculated fields, 133-137
Tableau Cloud, x
    Admin Insights project, 68
    authorizing Slack to use, 80
    integrating Pulse with Slack using, 106-110
    publishing data source to, 24
    relationship with Einstein Trust Layer,
        LMMs and, 9
    Settings section, 16
Tableau Data Catalog, 144
Tableau Desktop, viii
    constructing published data source using,
        20-25
    data professional's responsibilities when
        using, 9
Tableau Einstein, 147
Tableau Embedding API v3, 120
Tableau Host Mapping entry, 114
Tableau Lightning Web Components (LWCs),
    114
Tableau Mobile app, xi, 82-84
Tableau Prep Builder, 142-144
Tableau Pulse
    advanced metric definition
        calculated field as measure, 55-58
        custom definition filter, 59-64
        overview of, 55

definition filters, 37
    applying, 44-45
    defined, 43
    use case for, 43
embedding into other applications
    custom web pages or applications,
        120-121
    overview of, 112
    Salesforce CRM, 112-119
enabling, 16
future of, 148
gaps in, 148
insights in, 76-77
integrating with Slack
    overview of, 105
    Slack administration, 110-111
    Tableau Cloud administration, 106-110
LLMs in, 3
metric definitions in, 18, 36-41
    configuration options, 29
    Definition menu, 36
    insight types for, 40
    Insight window, 38-41
    number formats for, 38
    Options section, 38
    viewing, 71-73
metrics in, 19
    accessing, 33
    creating, by end users, 73-75
    digests of, 77-82
    followers for, 32
    interacting with, 34-36
    managing, 64-69
    sections of, 34
    selecting followers for, 32
    tracking usage, 68-69
    viewing, 71-73
metrics in, creating
    building metric definition, 26-30
    building published data source, 20-25
overview of, x
purpose of, 15
time dimension, 19, 29, 37, 148
    date offset setting, 54
    fiscal calendars, 51-53
    overview of, 46
    time granularity, 48-51
use cases
    finance, 93-97

About the Author

Ann Jackson is founder and managing director at Jackson Two, a boutique consulting firm specializing in AI and data analytics. In her practice, she empowers businesses to fully utilize their data assets. Ann has been recognized as a Tableau Visionary for five consecutive years, earning her the honorary title of Hall of Fame Visionary. She is hailed as an expert because of her proficiency with the product, her participation and leadership within the global community, and her contagious passion for data analytics. She's also coauthor of *Tableau Strategies*, the definitive book of real-world use cases using Tableau.

Colophon

The animal on the cover of *Learning AI Tools in Tableau* is the wire-crested thorntail hummingbird (*Discosura popelairii*). This species of hummingbird can be found in parts of Colombia, Ecuador, and Peru. There have also been sightings of this hummingbird in Bolivia. Because of the deforestation in the Amazon Basin, the wire-crested thorntail is listed as "Near Threatened."

The wire-crested thorntail hummingbird is one of the smallest birds in the world, measuring up to 4.5 inches in length and having an average weight of 2.5 grams. It is known for its unique and dramatic tail feathers, which consist of long, wire-like extensions that give it its name. The body of the wire-crested thorntail is a vibrant coppery green color with a white band above the tail. What makes this hummingbird most striking is the glossy, metallic green and bronze sheen on its face and throat, almost making the bird look like it glitters.

This hummingbird is typically found in humid forest and montane regions, where it feeds on nectar from a variety of flowers by using its specialized long, slender beak. It is also known to be territorial and may fiercely guard its feeding grounds.

Many of the animals on O'Reilly covers are endangered; all of them are important to the world.

The cover image is based on an antique line engraving from Wood's *Illustrated Natural History*. The series design is by Edie Freedman, Ellie Volckhausen, and Karen Montgomery. The cover fonts are Gilroy Semibold and Guardian Sans. The text font is Adobe Minion Pro; the heading font is Adobe Myriad Condensed; and the code font is Dalton Maag's Ubuntu Mono.

O'REILLY®

Learn from experts. Become one yourself.

60,000+ titles | Live events with experts | Role-based courses
Interactive learning | Certification preparation

Try the O'Reilly learning platform free for 10 days.